T0344441

POLYVINYLCHLORIDE

Environmental Aspects of a Common Plastic

POLYVINYLCHLORIDE

Environmental Aspects of a Common Plastic

WALTER TÖTSCH

and

HANS GAENSSLEN

Fraunhofer Institute for Systems Technology and Innovation Research

with the assistance of
Magdalena Sordo

ELSEVIER APPLIED SCIENCE
LONDON and NEW YORK

ELSEVIER SCIENCE PUBLISHERS LTD
Crown House, Linton Road, Barking, Essex IG11 8JU, England

Sole Distributor in the USA and Canada
ELSEVIER SCIENCE PUBLISHING CO., INC.
655 Avenue of the Americas, New York, NY 10010, USA

WITH 60 TABLES AND 18 ILLUSTRATIONS

ENGLISH LANGUAGE EDITION
© 1992 ELSEVIER SCIENCE PUBLISHERS LTD

This is the English-language version of
Polyvinylchlorid : zur Umweltrelevanz
eines Standardkunststoffes

© by Verlag TÜV Rheinland GmbH, Köln 1990 Gesamtherstellung:
Verlag TÜV Rheinland GmbH, Köln.

British Library Cataloguing in Publication Data applied for

ISBN 1-85166-774-1

Library of Congress CIP data applied for

v

Preface

This paper combines data on production, on processing and formulating, on application, on the waste stream and on the possibilities for recycling polyvinyl chloride insofar as such data has relevance for an assessment of environmental impact. It is intended to help place the PVC debate on a factually well-founded basis. The paper describes many, but not all facets of the environmental effects of a common plastic.

This book is based on work carried out at the Fraunhofer Institute for Systems Technology and Innovations Research and particularly on a report drawn up on the order of the Research Centre Jülich (Gaensslen, H., Sordo, M., Tötsch, W.: Production, Processing and Recycling of PVC. Order Number 011/41072711/930, April 1989). We would like to express our thanks to Dr. Kollmann of KFA Jülich for placement of this order. This book would not have come into being but for the assistance given by many colleagues. Magdalena Sordo carried out valuable preliminary work which forms the basis of many parts of the book. We would like to thank the following as representatives of our other colleagues: Eberhard Böhm for proof-reading, Günther Heger for the data base researches and Joachim Waibel for producing the illustrations.

The book has been translated by H.P. Kaufmann, Technical Translations, Marketing & Advisory Services, London. We are especially grateful to Harold M. Clayton and his colleagues at Hydro Polymers Ltd for proof-reading the English manuscript.

Contents

List of tables

List of figures

1. Introduction

1.1 Production and retention within the country

Polyvinyl chloride is the oldest thermoplastic and after polyethylene still constitutes the most important of the common plastics. The quantity of PVC produced in the Federal Republic of Germany (FRG) in 1987 was 1,320 kt [1]. Exports of PVC exceeded imports by around 243 kt. If one also takes into account the export trade balance of compounded PVC [2], there remain 1,063 kt PVC which were converted into semi-finished stock or finished products in Germany. On subtracting the export trade surplus for PVC semi-finished goods, the quantity remaining is 907 kt (always calculated as polymer). This calculation is shown in table 1 for the years 1975 to 1989. The export trade balance for PVC in finished products cannot be obtained from official statistics, either because PVC forms only a small part of the weight of the finished product - e.g. PVC parts in motor vehicles - or export trade statistics do not differentiate according to base materials. The FRG has a very large export trade surplus in converted products. The net export of PVC in finished products is hence likely to be positive. The amount of PVC retained within the country is therefore estimated to be about 800 kt.

TABLE 1.1

Production of PVC, conversion and consumption of semi-finished PVC goods in the FRG
[3]

Year	Raw Material production t	Conversion		Consumption of semi-finished goods	
		t	% of Prod.	t	% of Prod.
1976	965,076	913,719	95	845,690	88
1977	897,433	902,294	101	826,444	92
1978	1,006,265	936,785	93	851,221	85
1979	1,084,804	1,025,574	95	934,579	86
1980	953,189	923,823	97	843,805	89
1981	918,593	885,601	96	799,666	87
1982	864,372	824,890	95	721,957	84
1983	1,089,856	971,568	89	862,777	79
1984	1,131,926	960,108	85	831,999	74
1985	1,208,314	1,014,289	84	871,047	72
1986	1,241,865	1,047,200	84	894,620	72
1987	1,319,838	1,062,963	81	907,485	69
1988	1,411,513	1,146,792	81	1,033,705	73
1989	1,339,785	1,141,597	85	1,013,277	76

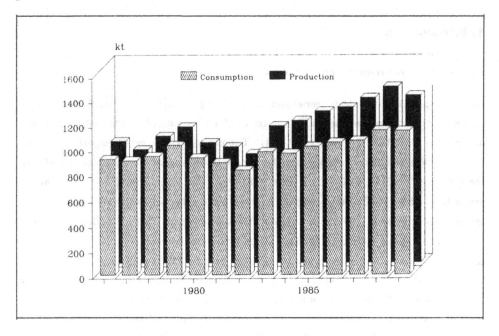

Figure 1.1:   Production and consumption of PVC compounds in the FRG [1,2]

1.2 Product range

Polyvinyl chloride is compatible with a whole series of additives. This gives the material an exceptionally wide product range. It extends from window profiles to pastes for fabric coating and has hardly changed since the beginning of the 80's in the Federal Republic of Germany. The quantity of rigid PVC (PVC-U) processed is about twice that of flexible PVC (PVC-P). This was not always so: At the beginning of the 70's it was PVC-P which predominated [4]. The PVC product range in the FRG differs from the West-European average on several points (Tables 1.2 and 1.3). Historically window profile manufacture played a far larger role in the Federal Republic than in other European countries. However this sector has also shown strong growth in England and France. In contrast, injection moulded articles, mainly shoes, are produced on a larger scale in Italy and France. Blow moulded articles from PVC - e.g. mineral water bottles - have captured a quarter of the market in France, while in the Federal Republic they do not even account for 2 % of the market. On the other hand, rigid PVC films account for 15 % of total consumption in the Federal Republic, while they are hardly manufactured at all in France and Austria.

TABLE 1.2
PVC consumption in Western Europe according to type of application

Product	1980 [4]		1985 [5]		1986 [3]	
	%	kt	%	kt	%	kt
Flexible PVC	**38.4**	**1,425.0**	**38.5**	**1,559.3**	**36.0**	**1,526.4**
Wire insulation	9.8	365.0	9.5	384.8	9.0	381.6
Flexible film	9.0	335.0	8.2	332.1	8.0	339.2
Floor coverings	5.4	200.0	5.1	206.6	5.0	212.0
Profiles, hoses	4.0	150.0	4.0	162.0	4.0	169.6
Coatings, pastes	6.7	250.0	3.9	158.0	7.0	296.8
Other	3.4	125.0	7.8	315.9	3.0	127.2
Rigid PVC	**61.6**	**2,290.0**	**61.5**	**2,490.8**	**64.0**	**2,713.6**
Films, slabs, sheets	9.7	360.0	10.5	425.3	12.0	508.8
Window profiles	4.0	150.0	12.9	522.5	5.0	212.0
Other profiles	10.0	370.0			8.0	339.2
Pipes, guttering	27.9	1,035.0	23.9	968.0	27.0	1,144.8
Blow mouldings	6.7	250.0	8.3	336.2	9.0	381.6
Injection mouldings	0.4	15.0	2.0	81.0	2.0	84.8
Records	1.9	70.0	1.9	77.0	1.0	42.4
Other	1.1	40.0	2.0	81.0		
Total	**100.0**	**3,715.0**	**100.0**	**4,050.0**	**100.0**	**4,240.0**

TABLE 1.3
PVC consumption in the FRG according to type of application

Product	1980 [4]		1985 [6]		1986 [7]	
	%	kt	%	kt	%	kt
Flexible PVC	**32.6**	**355.0**	**34.0**	**365.5**	**32.1**	**361.0**
Wire insulation	8.3	90.0	9.0	96.8	8.5	96.0
Flexible film	7.8	85.0	10.0	107.5	7.6	86.0
Floor coverings	5.5	60.0	6.0	64.5	3.6	40.0
Profiles, hoses	3.7	40.0	3.0	32.3	3.4	38.0
Coatings, pastes	6.4	70.0	6.0	64.5	7.3	82.0
Other	0.9	10.0	1.7	19.0		
Rigid PVC	**67.4**	**735.0**	**66.0**	**709.5**	**67.9**	**764.0**
Films, slabs, sheets	15.1	165.0	17.0	182.8	20.2	227.0
Window profiles	11.9	130.0	24.0	258.0	11.9	134.0
Other profiles	11.9	130.0			11.9	134.0
Pipes, guttering	22.9	250.0	20.0	215.0	16.8	189.0
Blow mouldings	2.3	25.0	3.0	32.3	2.2	25.0
Injection mouldings	0.5	5.0			1.6	18.0
Records	2.3	25.0	2.0	21.5	2.4	27.0
Other	0.5	5.0			0.9	10.0
Total	**100.0**	**1,090.0**	**100.0**	**1,075.0**	**100.0**	**1,125.0**

4

References

1. Produktion im produzierenden Gewerbe des In- und Auslands. Fachserie 4, Reihe 3.1, Published by: Statistisches Bundesamt (Federal Statistical Office), Wiesbaden, 1975 - 1989.

2. Außenhandel nach Waren und Ländern (Spezialhandel). Fachserie 7, Reihe 2 Published by: Statistisches Bundesamt, Wiesbaden, 1975 - 1989.

3. Gaensslen, H., Sordo, M., Tötsch, W., Produktion, Verarbeitung und Recycling von PVC, Fraunhofer-Institut für Systemtechnik und Innovationsforschung, Karlsruhe, April 1989.

4. Finkmann, H.-U., Einführung in die Verarbeitung von PVC. In Kunststoffhandbuch, Vol. 2 Polyvinylchlorid, Carl Hanser Verlag, München Wien, 1986, pp. 870 - 876.

5. Glenz, W., Der Kunststoffmarkt in Westeuropa. Kunststoffe 76, 1986, 834 - 877.

6. Data issued by the Association of Plastics Manufacturing Industry (Verband Kunststofferzeugende Industrie e.V., VKE).

7. Böhm, E., Tötsch, W., Cadmiumsubstitution - Stand und Perspektiven, Verlag TÜV - Rheinland, Cologne, 1989.

2. Manufacture of polyvinyl chloride

2.1 Manufacture of chlorine

2.1.1 Production, plant capacities and by-products

The chlorine content of pure polyvinyl chloride is 56.77 %. In 1987 manufacture of PVC amounted to 1,320 kt [1], which corresponds to 749 kt chlorine. The primary chlorine carrier is elemental chlorine which is reacted with ethylene to form dichloroethane. On cracking dichloroethane to give vinyl chloride, hydrogen chloride is formed. This hydrogen chloride is reacted either with acetylene to form vinyl chloride, or with ethylene and air or oxygen to form dichloroethane. Hydrogen chloride occurs in enormous quantities as a by-product in numerous technically important syntheses. Very often it has to be removed at considerable cost. The manufacture of VCM can, in principle, also serve as a sink for hydrogen chloride from other sources. With the existing plant structure this practice is exercised only on a small scale.

These days chlorine is produced by electrolysis. Starting products include hydrochloric acid, potassium chloride and above all sodium chloride. In 1987 about 93 % of chlorine was produced by electrolysis of sodium chloride. There are three processes which economically, technically and ecologically compete with each other: the amalgam process, the diaphragm process and the membrane process. The processes differ by the way that chlorine, which is produced at the anode, is kept away from caustic soda solution and hydrogen, which are formed at the cathode. All processes produce caustic soda solution and hydrogen apart from chlorine. The quality and concentration of the caustic soda solution is process related. The hydrogen has a high purity. Whether it can be utilized for other processes, or has to be burned, depends on the situation in the plant.

In the amalgam process, sodium amalgam is formed at the cathode, which consists of mercury. The sodium amalgam is reacted with water in a separate decomposer to produce sodium hydroxide and hydrogen. The caustic soda solution has a high degree of purity and a concentration of 50 %. Therefore it does not need to be concentrated by evaporation. The brine is recycled. The spent brine is brought up to the required concentration by adding rock salt and fed back to the electrolytic cell. The brine must have a high degree of purity as otherwise the mercury will be contaminated. Emissions of mercury occur in this process.

In the diaphragm process a diaphragm separates the anode from the cathode compartment. A steady stream of electrolyte through the diaphragm effects the separation of chlorine from

6

TABLE 2.1

Production, import, export and consumption of chlorine in the FRG according to Federal Statistical Office data [1,2] (tonnes)

Year	Production	Import	Export	Consumption
1975	2,295,949	50,199	114,599	2,231,549
1976	2,808,669	73,426	81,623	2,800,472
1977	2,807,658	90,699	55,433	2,842,924
1978	3,010,860	91,926	60,306	3,042,480
1979	3,201,631	126,661	103,925	3,224,367
1980	2,996,572	76,636	102,688	2,970,520
1981	3,013,153	81,160	97,145	2,997,168
1982	2,842,262	56,144	83,822	2,814,584
1983	3,136,492	70,620	64,223	3,142,889
1984	3,425,511	65,455	56,946	3,434,020
1985	3,493,447	74,051	49,022	3,518,476
1986	3,426,202	77,648	37,128	3,466,722
1987	3,452,142	119,075	48,096	3,523,121
1988	3,500,312	145,687	75,910	3,570,089
1989	3,442,943	114,041	73,837	3,483,147

Goods production numbers: 4111 110 [1]
Goods export trade number: 2801 300 [2]

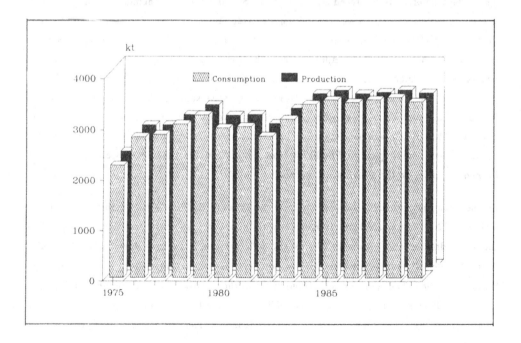

Figure 2.1:   Production and consumption of chlorine in the FRG [1,2]

the products formed at the cathode. The starting product may be brine which is cheaper than solid rock salt. The process requires about 20 % less electrical energy than the amalgam process. However, the resultant caustic soda solution is contaminated with common salt and has to be concentrated from 12 % to 50 % by evaporation which greatly increases the total energy consumption.

The membrane process is the most modern process for chlor-alkali electrolysis. A membrane separates the electrode compartments. This membrane allows sodium ions to permeate, but prevents the diffusion of chloride and hydroxyl ions. In order to protect the expensive membrane the brine has to be virtually free from alkaline earth ions. The brine is recycled. The starting product therefore is rock salt, just as in the case of the amalgam process. Current consumption is around 25 % less than for the amalgam process. The caustic soda solution obtained has a concentration between 30 and 35 %. Thus relatively little steam is needed to turn it into a 50 % solution with a very high purity.

The energy consumption of the chlor-alkali electrolysis is made up of the current consumption of the electrolytic cells, the transformers and other requirements. The sum of these values for each of the processes may be gathered from tables 2.7, 2.8 and 2.9. In table 2.2 a comparison is given between the current consumption of the electrolytic cells and the amount of energy necessary to evaporate the caustic soda solution to a commercially viable 50 % concentrate.

TABLE 2.2
Energy requirements of electrolytic cells and for the evaporation of caustic soda solution
(energy equivalent of steam) in kWh/t chlorine [3]

Process	Electrolytic cell	Evaporation
Amalgam process	3080 - 3400	0
Diaphragm process	2400 - 2700	800 - 1000
Membrane process	2200 - 2500	100 - 200

Chlorine production in the Federal Republic of Germany

The amalgam process remains economically the most important one. The membrane process is not currently used in Germany. It offers considerable advantages though and thus will be employed for all new plants. The relative importance of the processes is shown in table 2.3. Data have been extracted from various bibliographical sources [4,5,6].

TABLE 2.3
Quantitative importance of the processes for chlorine production in the FRG (tonnes)

Process	Amalgam million t	%	Diaphragm million t	%	Others million t	%
1979	2.50	78.09	0.53	16.44	0.18	5.47
1980	2.40	80.09	0.42	13.86	0.18	6.05
1981	2.40	79.65	0.44	14.74	0.17	5.61
1982	2.10	73.88	0.58	20.49	0.16	5.63
1983	2.20	70.14	0.77	24.52	0.17	5.34
1984	2.20	64.22	1.00	29.21	0.22	6.57
1985	2.20	62.98	1.08	30.81	0.22	6.22

The chlorine production capacities in the FRG are shown in table 2.4. The table also includes plants for the melt flow electrolysis of NaCl and plants which start from HCl or KCl.

TABLE 2.4
Chlorine production capacities (FRG 1989)

Company	Location	Capacity t/a	Process
BASF	Ludwigshafen	375,000	mercury/diaphragm
Bayer	Leverkusen	300,000	mercury/HCl
	Dormagen	330,000	mercury
	Brunsbüttel	50,000	HCl-electrolysis
	Uerdingen	140,000	mercury
	Leverkusen	20,000	HCl-electrolysis
	Dormagen	30,000	HCl-electrolysis
Hüls AG	Rheinfelden	60,000	mercury/diaphragm
	Marl	170,000	mercury
	Lülsdorf	80,000	mercury/KCl
Degussa	Knapsack	30,000	NaCl-melt flow
Dow	Stade	880,000	diaphragm
Elektrochemie	Ibbenbüren	145,000	mercury
Hoechst	Hoechst	320,000	mercury/diaphragm
	Gendorf	65,000	mercury
	Gersthofen	40,000	mercury
	Knapsack	85,000	mercury
ICI	Wilhelmshafen	130,000	mercury/diaphragm
Hoffmann-LaR.	Grenzach	10,000	mercury
Solvay	Rheinberg	290,000	diaphragm
Wacker	Burghausen	150,000	mercury
Total		**3,700,000**	
Production		**3,443,000**	
Yield (%)		**93**	

Availability of sodium chloride

About 78 % of the salt content of sea-water consists of sodium chloride. An even greater quantity of salt can be obtained from salt deposits in solid form. The German salt deposits alone are estimated to be 100,000 km^3 [7]. Rock salt contains impurities which would interfere with electrolysis. The largest part of these impurities are separated by precipitation with sodium hydroxide, sodium carbonate and barium hydroxide. The precipitate consists mainly of sulphates, carbonates and hydroxides of magnesium, calcium and barium. The quantity of separated residual substances depends on the quality of the salt which is used. The brine can be further purified with ion exchangers.

Sodium hydroxide as associated reaction product

Nowadays sodium hydroxide is manufactured exclusively by chlor-alkali electrolysis. In the Federal Republic more sodium hydroxide is produced than consumed. This excess is readily absorbed by foreign markets. The amount of caustic soda solution on offer depends on the requirement for chlorine. This accounts for the volatility of the price of NaOH. As recently as 1986 the price of sodium hydroxide was 30 dollar/tonne [8]. During this time sodium hydroxide displaced soda from many applications. Recently the demand for sodium hydroxide has grown strongly and its price rose correspondingly. In the autumn of 1988 up to 600 dollar per tonne of sodium hydroxide were paid in individual spot markets [8,9]. In the future too, the demand for caustic soda is expected to increase more strongly than the demand for chlorine [10].

From a technical aspect there is only one alternative to the manufacture of caustic soda solution by chlor-alkali electrolysis. This is the caustification of sodium carbonate. However, this process has not been used in industrialized countries for some time. Sodium carbonate is reacted with calcium hydroxide to form caustic soda and calcium carbonate.

$$Na_2CO_3 + Ca(OH)_2 \ = \ 2 \ NaOH + CaCO_3$$

Sodium carbonate is made from sodium chloride. The associated reaction product is calcium chloride and this can be only partly utilized.

$$2 \ NaCl + CaCO_3 \ = \ Na_2CO_3 + CaCl_2$$

TABLE 2.5

Production, import, export and consumption of sodium hydroxide in the FRG according to
Federal Statistical Office data [1,2] (tonnes)

Year	Production	Import	Export	Consumption
1975	2,489,035	18,141	515,544	1,991,632
1976	3,090,028	54,113	466,242	2,677,899
1977	3,081,167	91,780	598,679	2,574,268
1978	3,259,759	112,428	641,967	2,730,220
1979	3,414,913	93,742	761,199	2,747,456
1980	3,176,599	66,300	825,615	2,417,284
1981	3,209,174	71,753	818,506	2,462,421
1982	3,026,663	95,058	621,705	2,500,016
1983	3,349,995	117,567	720,519	2,747,043
1984	3,611,358	151,309	821,341	2,941,326
1985	3,696,749	145,440	917,351	2,924,838
1986	3,624,912	285,155	862,844	3,047,223
1987	3,635,873	152,010	897,748	2,890,135
1988	3,664,290	165,517	935,984	2,893,823
1989	3,541,102	200,043	807,751	2,933,394

Goods production number: 4146 110 and 4146 150 [1]
Goods export trade number: 2817 110 and 2817 150 [2]

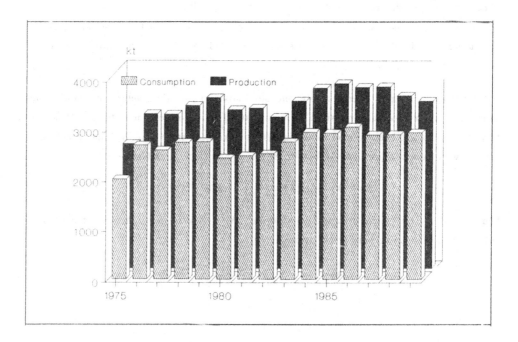

Figure 2.2: Production and consumption of sodium hydroxide in the FRG [1,2]

The largest part of the calcium chloride has to be disposed of as waste or discharged into the effluent. Both the manufacture of sodium carbonate and the caustification have a relatively high energy requirement. From an ecological point of view the manufacture of caustic soda solution by the sodium carbonate route offers no advantages in relation to chlor-alkali electrolysis.

TABLE 2.6
Use of sodium hydroxide in industrialized countries [8]

Field	%
Inorganic chemicals	20
Organic chemicals	17
Soap, detergents	4
Paper industry	14
Aluminium oxide	6
Rayon, cellophane	4
Neutralisation	12
Water treatment	1
Remainder	22

2.1.2 Chlorine production by the amalgam process

In the amalgam process the sodium chloride brine is recycled. After the dissolved chlorine has been removed from the brine, the latter is saturated with solid rock salt. Impurities are then precipitated by adding sodium carbonate, sodium hydroxide and barium carbonate. The NaCl solution has to meet higher purity demands than for the membrane and diaphragm processes because traces of heavy metals lower the hydrogen overvoltage at the mercury cathode.

The cathode consists of liquid mercury. During electrolysis metallic sodium is formed. This dissolves in the mercury as sodium amalgam. According to thermodynamic requirements the amalgam should react with water, but due to the high hydrogen overvoltage at the mercury electrode, this reaction does not take place. The sodium amalgam is pumped into a decomposer which, in principle, is a shorted out electrolytic cell. The decomposition catalyst consists of graphite activated with metal oxides. In this vessel sodium reacts with demineralized water to form a sodium hydroxide solution and hydrogen. The caustic soda solution has a concentration of 50 %. After removing of mercury traces, the purity is excellent.

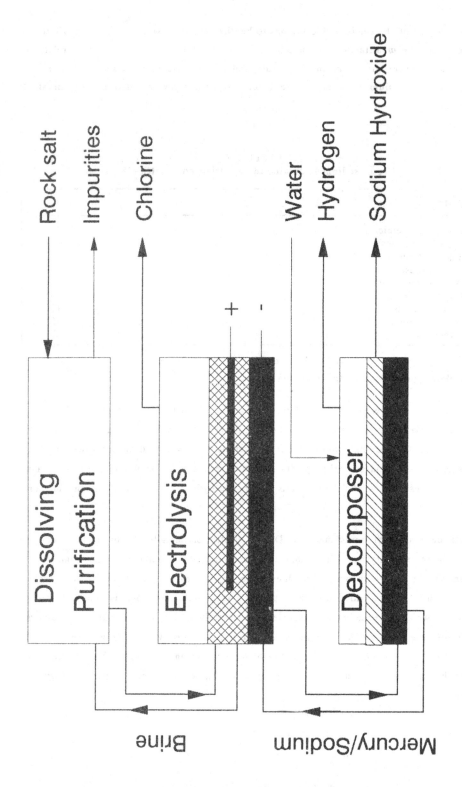

Figure 2.3: Amalgam process: Flow diagram

TABLE 2.7
Material balance and utilities requirement in the production of chlorine by the amalgam
process (related to 1 t chlorine) [11]

Input	
Rock salt	1,650 - 1,750 kg
NaOH, Na_2CO_3, $BaCO_3$, H_2O	50 kg
Chemically consumed water	508 kg
Water for solution	1,147 kg
Total	3,350 - 3,450 kg
Output	
Chlorine	1,000 kg
Sodium hydroxide 50 % (1120 kg dry)	2,240 kg
Impurities, losses (moist)	100 - 200 kg
Hydrogen	28 kg
Total	3,350 - 3,450 kg
Utilities requirement	
Electrical energy	3,600 - 3,800 kWh
Cooling water	100,000 kg
Process water	1,650 kg
Low-pressure steam	250 kg

The anode is dimensionally stable and consists of noble-metals and their oxides which have
been chemically precipitated on to titanium. The chlorine formed is cooled and dried with
concentrated sulphuric acid.

2.1.3 Chlorine production by the diaphragm process

In the diaphragm process the brine is not recycled but flows directly through the cell. The
starting product is rock salt or brine directly from the salt mine. Demands made on purity
are less severe than in the case of the other processes. Impurities such as iron, calcium or
magnesium, which could impair the functioning of the diaphragm, are removed by
precipitation with caustic soda or sodium carbonate. Sodium chloride, which is deposited
during concentration of the caustic soda solution, is fed back into the process.

TABLE 2.8

Material balance and utilities requirement in the production of chlorine by the diaphragm process (related to 1 t chlorine) [11]

Input	
Rock salt	1,650 - 1,750 kg
NaOH, Na_2CO_3	40 kg
Chemically consumed water	508 kg
Water for solution	1,147 kg
Total	3,350 - 3,450 kg
Output	
Chlorine	1,000 kg
Sodium hydroxide 50 % (1120 kg dry)	2,240 kg
Impurities, waste (moist)	100 - 200 kg
Hydrogen	28 kg
Total	3,350 - 3,450 kg
Utilities requirement	
Electrical energy	3,050 kWh
Cooling water	290,000 kg
Process water	4,300 kg
Low pressure steam	2,800 kg

Cathode and anode compartments are separated by a diaphragm. The NaCl solution is first pumped into the anode compartment where chlorine is formed. Chlorine is taken off, cooled, dried with sulphuric acid and compressed in the usual way. The electrolyte now flows through the diaphragm to the cathode. This flow is continuously maintained by a difference in level between the electrode compartments. At the cathode hydrogen evolves and caustic soda solution is formed. The stream of liquid through the diaphragm prevents the caustic soda diffusing back into the anode compartment. All diaphragm cells produce an alkaline brine containing 11 % sodium hydroxide and 18 % sodium chloride. Sodium chloride is precipitated by evaporation and cooling of the solution. The caustic soda solution obtained has a concentration of 50 % and still contains about 1 % sodium chloride. This purity is sufficient for most purposes, not however for the manufacture of cellulose fibres.

The diaphragm consists of 75 % asbestos. The asbestos is mixed with a polymer in order to meet the demands for mechanical strength. The manufacturers have no difficulties fulfilling the regulations regarding asbestos emissions. However, as waste disposal of used diaphragms becomes ever more expensive, experiments are being carried out with synthetic inorganic fibres as a replacement for the carcinogenic asbestos [3]. On average a diaphragm has to be replaced every two years.

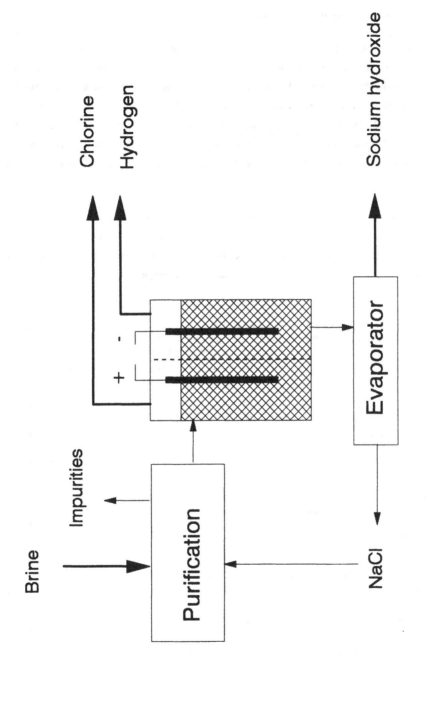

Figure 2.4: Diaphragm process: Flow diagram

2.1.4 Chlorine production by the membrane process

In this process a cation-exchange membrane separates anode and cathode compartments. The membrane is permeable to sodium ions, but to a large extent prevents the diffusion of chloride and hydroxide ions. The salt solution is first pumped into the anode compartment where chlorine is evolved at the electrode. The sodium ions migrate into the cathode compartment. Here hydrogen is evolved at the cathode and a 32 to 35 % solution of caustic soda is formed. This solution requires only a small amount of concentrating to achieve the standard commercial strength of 50 %. The chloride content of this caustic lye is less than 100 ppm.

A weakened brine is pumped out from the anode compartment. This brine is again saturated with NaCl and is recycled just as in the case of the amalgam process. In order to protect the costly membranes, the total concentration of calcium and magnesium ions has to be kept below 0.1 ppm. This is achieved by precipitation and ion exchange.

TABLE 2.9

Material balance and utilities requirement in the production of chlorine by the membrane process (related to 1 t chlorine) [11]

Input	
Rock salt	1,650 - 1,750 kg
NaOH, Na_2CO_3, $BaCO_3$, H_2O	50 kg
Chemically consumed water	508 kg
Water for solution	1,147 kg
Total	3,350 - 3,450 kg
Output	
Chlorine	1,000 kg
Sodium hydroxide 50 % (dry 1,120 kg)	2,240 kg
Impurities, losses (moist)	100 - 200 kg
Hydrogen	28 kg
Total	3,350 - 3,450 kg
Utilities requirement	
Electrical energy	2,800 kWh
Cooling water	100,000 kg
Process water	1,500 kg
Low pressure steam	800 kg

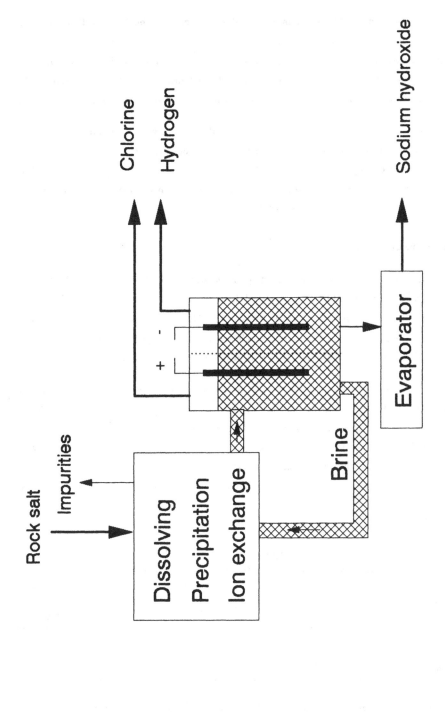

Figure 2.5: Membrane process: Flow diagram

It was the introduction of Nafion membranes [3] at the end of the 70's by DuPont which made possible the construction of membrane cells. The membranes have to be stable to concentrated alkali, display good conductivity and possess a high selectivity for sodium ions. Only perfluorinated polymers withstand the attack of concentrated caustic soda solution. The polymers used are perfluorinated polyethers with terminal sulphonate and carboxylate groups.

The membrane process combines the advantages of the amalgam and diaphragm processes. The caustic soda solution produced has a relatively high concentration and a high degree of purity. The electric energy consumption is 75 % of the amalgam process. Mercury emissions cannot occur. For these reasons a whole series of plants based on the membrane process have been built over the last few years, all of them outside the FRG.

2.2 Manufacture of ethylene

Ethylene is the most extensively produced petrochemical. In 1987 some 2,781 kt were produced in the Federal Republic of Germany [1]. In the world as a whole 97 % of ethylene are manufactured by the pyrolysis of hydrocarbons. The feedstock used in the USA is ethane, while in Western-Europe it is overwhelmingly (80 %) naphtha [12]. In the Federal Republic of Germany 17 % of ethylene serve to provide the feedstock for VCM.

In order to crack the hydrocarbons, they are mixed with steam and preheated to between 500 and 650 °C. In the actual reactor the mixture is cracked at temperatures between 750 and 875 °C. The ethylene yield is 28 to 31 %. In addition higher olefins, diolefins and methane are produced. Downstream from the reactor, the mixture of gases is cooled as rapidly as possible. In the final stage the product stream is washed, dried and fractionated. The specific energy consumption depends on the feedstock. According to Grantom and Royer [12] it is approximately:

TABLE 2.10
Energy requirement of ethylene production [12]

Feedstock	Energy requirement in GJ/t
Ethane	14.3
Propane	16.7
Naphtha	20.9

A comparison of the energy requirement has to be expressed in relative terms. The processes differ in the availability of the feedstocks, cost of plant and equipment and above all in the value of the by-products. It is particularly in the cracking of naphtha that by-products (e.g. propylene) are formed for which there is a large-scale requirement.

2.3 Manufacture of acetylene

Acetylene may be manufactured by partial combustion, electrothermally or from calcium carbide [13]. Hüls AG, the only VCM manufacturer in the FRG using acetylene feedstock, produces acetylene electrothermally. The acetylene plant capacity is 120,000 t/year. 90,000 t of this are required for VCM production. In the Hüls process a stream of hydrocarbons is passed through an electric arc. The residence time in the arc is just a few milliseconds. In this time the hydrocarbons are cracked into acetylene, ethylene, hydrogen and soot. After leaving the electric arc furnace the cracked gases still have a temperature of 1800 °C. This heat is utilized for the manufacture of additional ethylene in a "Prequench". For this purpose steam and additional hydrocarbon is passed into the hot gas. This causes a lowering of the temperature to 1200 °C. As a result of this prequenching the yield of ethylene is increased without affecting the yield of acetylene and hydrogen. Finally, the hot gas is cooled as rapidly as possible. Data on the Hüls electric arc process are given in Table 2.11.

TABLE 2.11

Material balance and utilities requirement in the manufacture of acetylene by the Hüls electric arc process (related to 1 t acetylene) [13]

Input	
Hydrocarbons for electric arc	1.8 t
Hydrocarbons for prequench	0.7 t
Output	
Acetylene	1.00 t
Ethylene	0.42 t
Hydrogen (3300 m^3)	0.30 t
Soot	0.45 t
Aromatics	0.08 t
Residues	0.12 t
Fuel gas	0.12 t
Utilities requirement	
Energy for electric arc	9,800 kWh
Energy for gas cleaning	2,500 kWh

2.4 Manufacture of vinyl chloride (VCM)

2.4.1 Production and plant capacities

Vinyl chloride is currently manufactured either by the thermal cracking of dichloroethane, or by the addition of hydrogen chloride to acetylene. Dichloroethane is produced either by direct chlorination or by oxychlorination of ethylene. All other manufacturing methods, e.g. the electrolysis of brine in the presence of ethylene, oxychlorination by means of nitrosyl chloride, the sodamin process [14] or the chlorination of ethane [15], are without commercial significance.

TABLE 2.12
Capacities for the manufacture of VCM (FRG 1989)

Company	Location	Capacity t/a	Basis
BASF	Ludwigshafen	100,000	Ethylene
Hüls AG	Marl	350,000	Ethylene/acetylene
Hoechst	Gendorf	160,000	Ethylene
	Knapsack	100,000	Ethylene
ICI	Wilhelmshafen	300,000	Ethylene
Solvay	Rheinberg	230,000	Ethylene
Wacker	Burghausen	225,000	Ethylene
Total		1,465,000	
Production		1,444,000	
Yield (%)		98	

The acetylene route is the older of the two processes and stems from a time when acetylene was one of the most important raw materials of chemical industry. These days ethylene, the starting product for the dichloroethane route, is markedly more economic. That is why world-wide more than 90 % of VCM is manufactured by the dichloroethane route. However, in the Federal Republic of Germany there is still one plant in operation which has around 25 % of the German VCM capacity and combines the chlorination of ethylene with the acetylene process. But also Hüls AG will switch to the oxychlorination process in the near future.

TABLE 2.13
Production, import and export of vinyl chloride in the FRG according to Federal Statistical
Office data [1,2] (tonnes)

Year	Production	Import	Export	Consumption
1975	831,514	124,098	63,439	892,173
1976	990,374	207,257	82,752	1,114,879
1977	912,822	188,088	98,613	1,002,297
1978	1,101,118	194,464	86,955	1,208,627
1979	1,132,802	242,703	85,412	1,290,093
1980	1,021,697	196,221	107,637	1,110,281
1981	902,684	150,344	71,107	981,921
1982	768,306	170,469	118,670	820,105
1983	1,232,188	149,499	196,161	1,185,526
1984	1,280,254	120,323	205,890	1,194,687
1985	1,346,176	140,229	212,751	1,273,654
1986	1,292,002	161,986	182,645	1,271,343
1987	1,434,168	160,027	222,526	1,371,669
1988	1,458,874	177,629	202,774	1,433,729
1989	1,443,719	195,253	193,184	1,445,788

Goods production number: 4227 70 (Vinyl chloride, Vinylidene chloride) [1]
Goods export trade number: 2902 310 (Vinyl chloride) [2]
*Production figures also include vinylidene chloride (VDC)

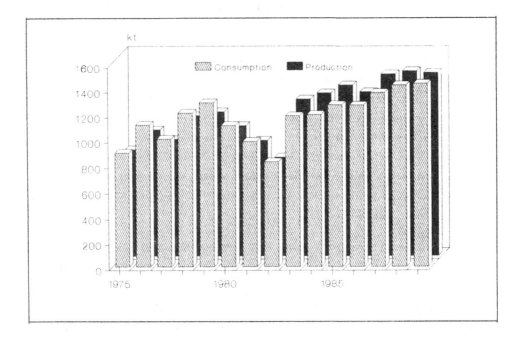

Figure 2.6:   Production and consumption of vinyl chloride in the FRG [1,2]

2.4.2 Manufacture of vinyl chloride from ethylene and chlorine

These days, the largest part of vinyl chloride (VCM) production is carried out in plants which employ the following three processing steps:

1) Low temperature liquid phase chlorination of ethylene to 1,2-dichloroethane (EDC)

$$C_2H_4 + Cl_2 = C_2H_4Cl_2$$

2) Oxychlorination of ethylene in the gas phase at medium temperature to form EDC

$$C_2H_4 + 2\ HCl + 1/2\ O_2 = C_2H_4Cl_2 + H_2O$$

3) Pyrolysis of EDC at high temperature to VCM

$$C_2H_4Cl_2 = 2\ C_2H_3Cl + HCl$$

The overall reaction is:

$$2\ C_2H_4 + Cl_2 + 1/2\ O_2 = 2\ C_2H_3Cl + H_2O$$

This balanced process permits the HCl, formed during pyrolysis, to be fully employed as a raw material for EDC manufacture.

The direct chlorination of ethylene is a strongly exothermic process. By skilful process operation the heat of reaction can be utilized for the purification of EDC by distillation. Yield and selectivity are excellent: 99 % of ethylene react. Of the reaction products 99,8 % are EDC. The catalyst is $FeCl_3$. The oxychlorination reaction is carried out at 230 to 315 °C. The pressure is 3 to 13 bar depending on the process. Metal chlorides (e.g. copper chloride) serve as catalysts on inert supports. The oxychlorination reaction can be carried out with air or oxygen. These days slightly more than one third of the production capacity is based on pure oxygen. In both cases the reaction proceeds virtually to completion. EDC accounts for 96 to 98 % of final products.

EDC has to be distiled to obtain a purity of 99.5 % minimum thus removing water and impurities which could interfere in the pyrolysis step. The pyrolysis of EDC to VCM is carried out with high yields at temperatures between 500 and 600 °C and pressures of 10 to 35 bar. The residence time is between 5 and 20 seconds depending on temperature, the conversion per pass is 50 to 60 %. VCM selectivity reaches 98 to 99 %. Acetylene, benzene, various chlorinated hydrocarbons and tars occur as by-products. The VCM formed is purified by distillation to achieve the quality necessary for polymerization.

TABLE 2.14

Material balance and utilities requirement in the manufacture of vinyl chloride by the
ethylene/chlorine process (related to 1 t VCM) [11]

Input	
Ethylene	470 kg
Chlorine	580 kg
Oxygen	128 kg
Total	1,178 kg
Output	
Vinyl chloride	1,000 kg
Water	144 kg
By-products	34 kg
Total	1,178 kg
Utilities requirement	
Electrical energy	218 kWh
Cooling water	290,000 kg
High pressure steam	1,640 kg
Low pressure steam	900 kg
Boiler feed water	1,000 kg
Fuel	4,200 MJ

TABLE 2.15

Production, import and export of dichloroethane in the FRG according to Federal Statistical
Office data [1,2] (tonnes)

Year	Production	Import	Export	Consumption
1975	723,000	8,919	12,624	719,995
1976	1,030,833	6,036	13,337	1,023,533
1977	1,371,909	2,830	94,338	1,280,402
1978	1,320,950	3,271	108,779	1,215,442
1979	1,300,386	5,452	101,042	1,204,796
1980	1,194,621	1,485	118,674	1,077,432
1981	1,123,103	1,201	101,172	1,023,132
1982	964,290	22,341	156,108	830,523
1983	1,504,461	58,480	158,111	1,404,830
1984	1,683,052	89,017	139,951	1,632,118
1985	1,725,703	88,372	134,202	1,679,874
1986	1,648,357	61,639	167,394	1,542,602
1987	1,758,121	113,949	107,235	1,764,835
1988	1,634,875	158,417	136,904	1,656,388
1989	1,597,903	215,720	152,980	1,660,643

Goods production number: 4227 20 [1]
Goods export trade number: 2902 260 [2]

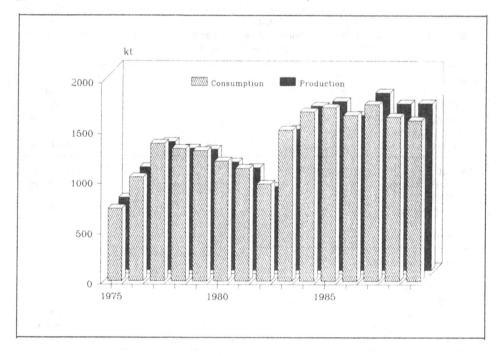

Figure 2.7: Production and consumption of dichloroethane in the FRG [1,2]

2.4.3 Manufacture of VCM from acetylene

In the oldest vinyl chloride process HCl adds on to acetylene:

$$C_2H_2 + HCl = C_2H_3Cl$$

HCl addition proceeds smoothly either under normal pressure or slight excess pressure, at temperatures between 150 and 250 °C. By using mercury (II) chloride on an activated carbon carrier as catalyst, high yields and very good selectivities are obtained. Tubular or fluidized bed reactors are used. Apart from unreacted acetylene and hydrogen chloride, the reaction gas contains VCM and small quantities of EDC and acetaldehyde. A water wash, to remove the largest part of the HCl, is followed by an alkaline wash. By compressing VCM to 7 to 8 bar and by cooling, the largest part of it can be taken off as a liquid. The crude VCM is decanted from water and freed from dissolved acetylene. Pure VCM is obtained by subsequent distillation, while the by-products (essentially EDC and acetaldehyde) remain as bottom products.

Acetylene is washed out of the waste gas streams with N-methyl pyrrolidone and returned to the process. The remaining waste gas stream is burned. The acetylene process produces markedly less chlorinated by-products than the oxychlorination. Mainly EDC is formed as a by-product by the addition of two molecules HCl to one molecule acetylene.

For economic reasons the acetylene process is not used alone. However, the combination of the processes for the manufacture of VCM from acetylene and from ethylene has obtained a limited importance. This involves two parallel production lines one concerned with the hydrochlorination of acetylene and the other with the chlorination of ethylene to EDC. In this procedure the HCl resulting from pyrolysis of EDC covers the HCl requirement for the acetylene line. In the Federal German Republic this process combination has been successfully operated by Hüls AG for a long time. Over the last few years this manufacturer produced substantially more VCM on acetylene basis (70 %) than on ethylene basis (30 %) in order to utilize surplus HCl accumulating from other reactions.

TABLE 2.16

Material balance and utilities requirement in the manufacture of vinyl chloride from acetylene and HCl (related to 1 t VCM) [11]

Input	
Acetylene	420 kg
HCl	600 kg
Water	2 kg
Total	1,022 kg
Output	
Vinyl chloride	1,000 kg
EDC	4 kg
Acetaldehyde	5 kg
Losses	13 kg
Total	1,022 kg
Utilities requirement	
Electrical energy	100 kWh
Cooling water	100,000 kg
Low pressure steam	1,200 kg
Medium pressure steam	300 kg

2.5 Polymerization of vinyl chloride

2.5.1 Production and plant capacities

The three processes to manufacture polyvinyl chloride are the suspension process, the mass process and the emulsion process. All are based on free-radical initiated polymerization of vinyl chloride. The processes differ in the form in which the VCM enters into the reaction. The manufacturing process affects the properties of the PVC and consequently its type of application (see chapter 4).

The suspension process is by far the most important one. VCM is stirred in water to produce a fine dispersion. The initiator, which most frequently is an organic peroxide, dissolves in the VCM droplets. The suspending agent prevents the agglomeration of the newly formed particles. The reaction parameters can be controlled exactly and product properties such as molecular weight can be varied within wide limits.

In the emulsion process an emulsifier is used to form an extremely fine dispersion of vinyl chloride in water. The initiator is dissolved in the water. Polymerization results in a fine dispersion of PVC in water which is subsequently precipitated and spray dried. Emulsion PVC is the preferred product for the manufacture of pastes.

The mass process involves the polymerization of pure VCM. As PVC is virtually insoluble in vinyl chloride, it precipitates. In terms of process technology the procedure is very simple. As no emulsifiers or suspending agents are used, the resultant product is very pure. A disadvantage is the relative inflexibility of the process. Copolymers cannot be produced by this mode.

DIN 7746 specifies that VC homopolymers may contain a maximum of 10 mg VCM/kg. This is achieved by steam stripping in a boiler or in a perforated plate column. If this limit value is adhered to, the PVC converter can be assured that the TRK-value (a limit concentration for carcinogens) for VCM will not be exceeded at the working place while processing PVC. Residual monomer contents below 1 ppm are the rule for many products [16]. Unconverted vinyl chloride is recovered from the autoclaves or the monomer stripping unit and recycled to the polymerization process. Waste gas streams with a low concentration of residuals are incinerated wherever this is possible.

TABLE 2.17
PVC production capacities (FRG 1989)

Company	Location	Capacity t/a	Process
BASF	Ludwigshafen	150,000	Suspension/Emulsion
Hüls AG	Marl	380,000	Suspension/Mass/Emulsion
Hoechst	Gendorf	150,000	Suspension/Mass/Emulsion
	Knapsack	95,000	Emulsion
EVC	Wilhelmshafen	190,000	Suspension
Solvay	Rheinberg	160,000	Suspension/Emulsion
Wacker	Burghausen	180,000	Suspension/Emulsion
	Cologne	170,000	Suspension/Mass
Total		1,475,000	
Production		1,340,000	
Rate of utilization (%)		90	

2.5.2 Suspension polymerization

In suspension polymerization the VCM is finely dispersed in water by vigorous stirring. Polymerization is started by addition of initiators which dissolve in the monomer. A suspension stabilizer forms a protective monomolecular layer on the surface of the particles thus ensuring that the droplets do not coalesce and that the PVC particles do not agglomerate.

TABLE 2.18
Example of a formulation for the polymerization of VCM [11]

Components	Quantity (kg)
VCM	100
Demineralized water	200
Methyl cellulose	0.05
Cumene peroxyneodecanoate	0.056
Bis(2-ethylhexyl)peroxydicarbonate	0.03

Polymerization is carried out discontinuously in vessels fitted with an agitator and with a cooling jacket and cooling elements to remove the heat of polymerization. The temperature is kept constant within a range of +/- 0.5 °C because it influences the molecular weight and hence the properties of the polymer. During the polymerization process, which lasts 6 to 8 hours, the content of the autoclave is vigorously stirred. When 90 % of the VCM have polymerized, the reaction is terminated, the pressure is released and the content of the stirred autoclaves is transferred to a receiver. In the process the PVC is degassed, i.e. the largest part of the unreacted VCM is removed by vacuum and recovered.

TABLE 2.19
Material balance and utilities requirement in the manufacture of PVC by the suspension process (related to 1 t PVC)

Input		
Vinyl chloride	1,010	kg
Methyl cellulose	0.5	kg
Organic peroxides	0.9	kg
Total	1,011.4	kg
Output		
PVC	1,000	kg
Losses in waste water	3	kg
Losses as PVC dust	1	kg
Other losses	7.4	kg
Total	1,011.4	kg
Utilities requirement		
Electrical energy	200	kWh
Cooling water	100,000	kg
Medium pressure steam	1,000	kg
Demineralized water	3,000	kg

Next steam is blown through the suspension to remove further VCM. These vapours too are transferred into the VCM recovery system. The steam-stripped suspension, which has a solids content of about 32 %, is first of all centrifuged to get rid of most of the water. The next stages are drying with warm air, storing in silos and packaging the finished product. Gas streams from the plant, whose VCM content can no longer be reduced economically, are passed through an incinerator in order to keep the VCM emissions as low as possible.

2.5.3 Emulsion polymerization

In this process VCM is emulsified in water with the aid of an emulsifier. Its task is to produce micelles in which polymerization takes place, to stabilize the monomer droplets in the emulsion and to stop the polymer particles from agglomerating. This type of polymerization gives rise to particularly fine particles which are specially suitable for the production of PVC pastes and dispersions.

TABLE 2.20

Material balance and utilities requirement in the manufacture of PVC by the emulsion process (related to 1 t PVC) [11]

Input		
Vinyl chloride	1,022	kg
Redox system, pH stabilizer, emulsifier	23	kg
Total	1,045	kg
Output		
PVC	1,000	kg
Redox system, pH stabilizer, emulsifier	23	kg
PVC losses (dust, oversize material)	14	kg
PVC losses in waste water	2	kg
PVC losses in waste gases	6	kg
Total	1,045	kg
Utilities requirement		
Electrical energy	500	kWh
Cooling water	250,000	kg
Low pressure steam	900	kg
Medium pressure steam	600	kg
Fuel	3,570	MJ
Demineralized water	3,000	kg

While the suspending agent plays a big part in suspension polymerization, it is the emulsifier which is of particular importance in emulsion polymerization. Peroxidic redox-systems act as polymerization catalysts. The polymerization is carried out in stirred autoclaves fitted with cooling jackets and cooling fins to remove the heat of polymerization. The temperature is kept constant with great accuracy because it determines the quality of the PVC. The polymerization process lasts 6 to 9 hours. During this time the autoclave content is constantly stirred. The reaction is terminated when about 90 % of the VCM have been converted. The largest part of the VCM is removed by flash distillation and steam stripping. The VCM containing vapours are fed to the VCM recovery system.

2.5.4 Mass polymerization

Polymerization of PVC in the mass is carried out in less complicated production plants because there is no aqueous phase. As neither an emulsifier nor a suspending agent is used, the products are purer than in the emulsion or suspension processes. The high purity of the PVC gives rise to very stable polymers with outstanding transparency. However, polymerization in bulk does throw up several problems. Firstly, it is more difficult to remove the heat of polymerization. Secondly, it is difficult to stir the polymerizate because the viscosity increases very sharply as polymerization progresses. Finally, the method is suitable only for the production of standard PVC, while the production of copolymers encounters difficulties. It is primarily this lack of flexibility why this process is only used on a limited scale.

VCM is fed into a prepolymerization reactor. Initiators are added and slight initial conversion takes place. The prepolymerizate is then transferred to the actual polymerization autoclave which is usually a horizontal stirred vessel. A screen-wiper type of stirrer, driven by a very powerful motor, operates inside the cooled reactor. The reaction is terminated when 75 to 80 % of the VCM have been converted to PVC. The largest part of the unconverted VCM is flashed off and recycled to the polymerization process. The residual VCM in the polymer particles is removed by repeatedly applying vacuum and pressurizing with nitrogen. The polymer particles are ground, screened and packaged.

TABLE 2.21

Material balance and utilities requirement in the manufacture of PVC by mass polymerization (related to 1 t PVC) [11]

Input		
Vinyl chloride	1,015	kg
Catalyst (DEPC)	0.4	kg
Total	1,015.4	kg
Output		
PVC	1,000	kg
Catalyst and its decomposition products	0.4	kg
VCM losses	10.1	kg
PVC losses	4.9	kg
Total	1,015.4	kg
Utilities requirement		
Electrical energy	327	kWh
Cooling water	66,000	kg
Low pressure steam	140	kg
Medium pressure steam	60	kg
High pressure steam	230	kg
Process water	180	kg

References

1. Produktion im produzierenden Gewerbe des In- und Auslands. Fachserie 4, Reihe 3.1, Published by: Statistisches Bundesamt, Wiesbaden, 1967 - 1989.

2. Außenhandel nach Waren und Ländern (Spezialhandel). Fachserie 7, Reihe 2, Published by: Statistisches Bundesamt, Wiesbaden, 1975 - 1989.

3. Schmittinger, P. et al., Chlorine. In Ullmann's Encyclopedia of Industrial Chemistry, Vol. A6, VCH Verlagsgesellschaft, Weinheim, 1986, pp. 399 - 477.

4. Rauhut, A., Quecksilberbilanz 1983 - 1985. Landesgewerbeanstalt Bayern, Nürnberg, 1988.

5. Rauhut, A., Quecksilberbilanz 1980 - 1982. Landesgewerbeanstalt Bayern, Nürnberg, 1985.

6. Rauhut, A., Quecksilberbilanz 1977 - 1979. Wasser Luft Betrieb 1982, 5, 50 - 51.

7. Gülpen, E. et al. Natriumchlorid. In Ullmann's Enzyklopädie der technischen Chemie, Vol. 17, VCH Verlagsgesellschaft, Weinheim, 1980, pp. 179 - 199.

8. Optimistic Outlook for Chloralkali Business. European Chemical News 1988, 1327/50, 9.

9. Caustic Soda Prices Set for Long Awaited Upturn. European Chemical News 1988, 1336/50, 6.

10. Slow Growth for US Chloralkali Market. European Chemical News 1989, 1357/52, 10.

11. Gaensslen, H., Sordo, M., Tötsch, W., Produktion, Verarbeitung und Recycling von PVC. Fraunhofer-Institut für Systemtechnik und Innovationsforschung, Karlsruhe, April 1989.

12. Grantom, E., Royer, D.J.: Ethylene. In Ullmann's Encyclopedia of Technical Chemistry, Vol. A10, VCH Verlagsgesellschaft, Weinheim, 1986, pp. 45 - 93.

13. Pässler, P. et al., Acetylene. In Ullmann's Encyclopedia of Technical Chemistry, Vol. A1, VCH Verlagsgesellschaft, Weinheim, 1986, pp. 97 - 145.

14. Van Andel, E., An electrolysis-free route to vinyl chloride and soda ash. Chem. Ind. 2, 1983, pp. 139 - 141.

15. Rossberg, M. et al., Chlorinated Hydrocarbons. In Ullmann's Encyclopedia of Technical Chemistry. Vol. A6, VCH Verlagsgesellschaft, Weinheim, 1986, pp. 233-399.

16. Baumgärtel, H.-G., Kunststoffe und Umweltschutz. Kunststoffberater 1985, 12,33 - 35.

3. Emissions caused by the manufacture of polyvinyl chloride

3.1 Mercury

In the manufacture of PVC, mercury emissions can occur in the production of chlorine by the amalgam process and in the synthesis of vinyl chloride from acetylene. The acetylene process employs mercury salts as catalyst. No data are available on the volume of mercury emissions from the acetylene process, although an upper limit can be given in the following terms: in 1985 the whole German chemical industry used 3.9 t mercury for catalyst purposes [1]. It is likely that the actual emissions would be much smaller. In chlorine production by the amalgam process, there are emissions of mercury into the products, the waste water and the exhaust air. These currently run at less than 3 g Hg/t Cl_2 (without landfill) [2]. Additionally, mercury finds its way to special waste landfills as impurity in poisoned catalysts, filter residues and replaced parts of plant.

Mercury emissions into the products

In the Federal Republic the total emissions of mercury into the products of chlor-alkali electrolysis amounted to 1.1 t in 1985 [1].

Chlorine: The chlorine contains only traces of Hg and these condense almost quantitatively on cooling. The mercury content of the dried chlorine amounts to 0.001 to 0.1 g/t chlorine depending on the operational care taken [2].

Hydrogen: The initially quite high mercury content is brought down to 0.002 - 0.015 mg/m^3 hydrogen by adsorption on activated carbon. The mercury content is reduced to below 0.001 mg/m^3 by adsorption on copper/aluminium oxide or silver/zinc oxide [2].

Caustic soda solution: By centrifuging or filtering through filters coated with activated carbon, the mercury concentration can be reduced to below 0.1 g/t [2].

Mercury emissions into waste water

As the amalgam process involves closed circuit brine recycling, emissions into the waste water can occur only through leaking seals, cleaning operations and condensates from gas purification. The quantity of waste water can be reduced to 0.3 to 1.0 m^3 per tonne chlorine produced. The mercury content of the waste water can be lowered by reduction, adsorption or liquid/liquid extraction. In the USA, the Clean Water Act has, since 1982, limited the residual mercury content to 0.1 g per tonne chlorine produced [2]. An EC directive has limited mercury emissions to 1 g per tonne chlorine since 1982 [3]. In 1985 the emissions into waste water were 0.2 tons in the Federal Republic, corresponding to 0.09 g per tonne chlorine [1].

Mercury emissions into exhaust air

The air in the cell rooms is changed approximately 15 - 20 times per hour. The German TA-Luft directive lays down a maximum emission of 1.5 g mercury per tonne chlorine for new plants and 2.0 g/t for plants authorized prior to 1972 [4]. In 1985 the emissions into the exhaust air amounted to 4.2 tonnes, corresponding to 1.9 g per tonne chlorine [1].

Mercury in landfills

In 1985, 36.3 tonnes mercury ended up on special landfills together with filter sludge, spent catalysts, residues from product purification and scrapped plant parts. This corresponded to 16 g per tonne chlorine [1]. It is possible to recover a part of the mercury by distillation in closed retorts.

TABLE 3.1

Mercury emissions in the production of chlorine according to Rauhut [1,5,6] (tonnes)

Year	Water	Air	Products	Waste tip	Total
1973	-	-	-	-	128.0
1974	-	-	-	-	127.0
1975	-	-	-	-	99.0
1976	-	-	-	-	103.0
1977	4.0	18.0	10.0	46.0	78.0
1978	2.5	20.0	7.5	57.0	87.5
1979	2.5	17.5	5.0	62.5	87.5
1980	2.4	14.4	2.4	52.8	72.0
1981	1.2	10.8	2.4	36.0	50.4
1982	1.1	9.5	2.1	33.6	46.3
1983	1.1	5.5	1.1	33.0	40.7
1984	0.4	4.8	1.1	25.3	31.6
1985	0.2	4.2	1.1	36.3	41.8
Specific Hg emissions					
g Hg/t Cl_2	0.09	1.91	0.5	16.5	19.0
g Hg/t PVC	0.05	1.08	0.3	9.4	10.8

In 1985, 2.2 million t chlorine were produced by the amalgam process. For every tonne of this chlorine 2.5 g mercury were emitted. When PVC is manufactured from amalgam produced chlorine, for every tonne of PVC less than 1.5 g Hg were emitted and about 10 g Hg were deposited in landfills.

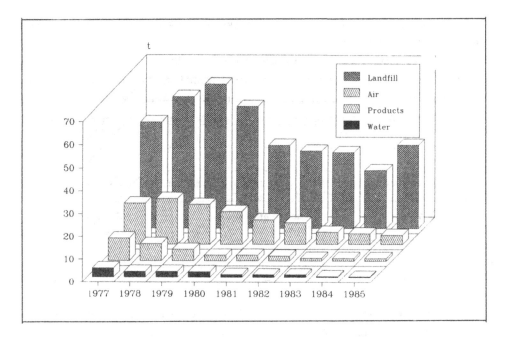

Figure 3.1:   Development of mercury emissions from chlorine production: FRG 1973-1985

3.2 Chlorine

Chlorine emissions occur in production. Apart from this, they can take place only due to leakages in storage or handling. Under TA-Luft 86, chlorine falls into Class II for vaporous and gaseous materials. Accordingly, the exhaust gas concentration may not exceed 5 mg/m^3, if the emission level is 50 g/h or over. Plants for the production of chlorine are specially regulated: The emissions may not exceed 1 mg/m^3, or 6 mg/m^3 for plants with full liquefaction [4]. The chlorine emissions would thus lie well below 1 g/t chlorine produced.

3.3 1,2-Dichloroethane (EDC)

1,2-dichloroethane (EDC) has recently been classified as carcinogenic in animal experiments (group III A2) [7]. As a result the MAK-value of 20 ml/m^3 or 80 mg/m^3 has been suspended. MAK-values (maximum allowable work place concentrations) are based on toxicologic considerations. The assumptions made do derive these values do not apply to carcinogens. The TRK-value, a limit value for carcinogens, is based on state of the art exposure minimisation technique. As yet, there is no TRK-value for EDC defined.

Emissions are regulated by TA-Luft. The vent gases must be fed to a waste gas cleaning plant. The EDC concentration in the vent gas must not exceed 5 mg/m^3 [4]. Emissions of EDC linked with PVC manufacture originate mainly from the oxychlorination process. The quantity of vent gas depends on the oxidizing agent. If air is used as oxidizing agent much more vent gas has to be cleaned than if the reaction is based on pure oxygen. Table 3.2 shows the amount of vent gas from these processes and the maximum allowable emissions.

TABLE 3.2
Quantity of vent gas from EDC manufacture and maximum allowable EDC emission

Oxidizing Agent	Quantity of vent gas Million m^3/a m^3/t EDC		Tolerated emission t/a g/t EDC	
Air	282	240	1.41	1.2
Oxygen	4.7	8	0.02	0.04

However, there are also emissions of less well defined sources, like vent gases from distillation, leaks etc. Actual EDC emissions will be between 10 and 30 t per annum. In the atmosphere EDC is decomposed more slowly than VCM. The half-life in the atmosphere is estimated to be 3 to 4 months [9,10]. However, the reliability of this half-life seems to be only moderate [11].

3.3 Vinyl chloride

The danger of vinyl chloride was underestimated for a long time because of its low acute toxicity. It was only at the beginning of the seventies that an accumulation of observations indicated that VCM causes a rare form of liver cancer - an angiosarcoma. The MAK value, which in 1966 was fixed at 500 ppm, was therefore withdrawn. VCM is now clearly identified as a carcinogenic substance (Group III A1 of Gefahrstoffverordnung). The TRK-value is 3 ppm or 8 mg/m^3 for older plants and 2 ppm or 5 mg/m^3 for new VCM or PVC manufacturing plants [7]. Under the TA-Luft directive VCM has been placed in Class III for cancer-promoting substances: Exhaust gases may not exceed a concentration of 5 mg VCM [m^3, if the mass flow is greater than 25 g/h.

Vinyl chloride emissions can result from the manufacture of VCM, the polymerization to PVC and, in theory, from processing, utilisation or destruction. The emissions from manufacture of VCM (20 - 30 t/a) and in the processing of polyvinyl chloride (<5 t/a) are very low. Emissions arising during use and destruction of PVC are negligible for PVC does not depolymerize. Large quantities of air have to be used to strip the polymer from residual monomer in order to obtain a safe product. Such large quantities of air cannot be completely purified by current state of the art technology. Annual emission of approximately 300 t VCM arise from monomer stripping [8]. Vinyl chloride has a half-life of 2.4 days in the atmosphere. It is decomposed by reaction with OH radicals [8].

VCM emissions from VCM production

As recently as 1974 the specific emissions of VCM were 0.5 kg per tonne of VCM produced. Until 1976 this emission could be reduced by around 95 % [12]. Nowadays VCM-emissions from manufacture are estimated to be 20 to 30 t/a [8] corresponding to 1.3 to 2.0 g/t VCM.

VCM emissions from polymerization

The ecological aim in the polymerization of VCM is to obtain a polymerizate with as low a monomer content as possible and to keep the process emissions low. The actual polymerization is carried out in a closed system. Exhaust gas streams containing VCM originate from the removal of inert gas streams from the autoclaves [13]. After polymerization the polymer is subjected to an intensive degassing which is also carried out in the closed system. Notwithstanding this intensive degassing the polymer still contains VCM. The TA-Luft requires that the residual content of vinyl chloride be kept as low as possible before the transfer from the closed system to the drying stage. The average monthly levels may not exceed the following limit values for VCM content:

TABLE 3.3
Maximum values for the VCM content of PVC before the transfer from the closed system according to the TA-Luft directive (monthly average)

PVC type	Maximum VCM content g VCM/t PVC
Mass PVC	10
Suspension homopolymer	100
Suspension copolymer	400
Micro-suspension PVC	1500
Emulsion PVC	1500

After the transfer into the open system, the polymer is dried with very large quantities of air and is stripped of monomer. It is quite usual to use drying air quantities up to 60,000 m^3 per tonne PVC [13]. To clean these quantities of air is very costly. For this reason TA-Luft sets aside the limit value for VCM emissions from PVC manufacturing plants. From production quantities of each type of PVC [13] and the permitted residual contents on transfer from the closed system, it is possible to derive upper limits for the VCM emission from PVC plants:

TABLE 3.4
Production quantity [13] and maximum VCM emissions from various types of PVC (year of reference: 1978)

Type of PVC	Production quantity t	VCM emission t
Mass PVC	95,000	1
Suspension homopolymer	505,000	51
Suspension copolymer	125,000	50
Micro-suspension PVC	85,000	128
Emulsion PVC	210,000	315
Total	**1,020,000**	**544**

Given the technology of 1978 the corresponding emission would be 700 t in 1989. However, the residual VCM concentrations on transfer from the closed system have been lowered considerably since 1978. Also, TA-Luft directs that the vent air from the drier be used to the greatest possible extent as combustion air in furnaces. By employing combustion the limit value of 5 mg VCM [m^3 vent air can certainly be met [14]. Therefore VCM emissions from polymerization plants probably did not exceed 300 t during 1988 [8].

3.4 Emissions into effluent

In the oxychlorination process effluents occur as reaction water, water introduced with combustion air, EDC washing water, and steam stripping condensate. The specific quantity of effluent is given as 0.4 m^3/t VCM [12]. The effluent streams are neutralized, steam stripped and subjected to biological cleaning [15]. VCM loaded effluents also occur in suspension and emulsion polymerization. They are treated by means of phase separators, biological wastewater treatment plants, stripping or adsorption processes used alone or in combination [15]. In 1990 a German plant with a production capacity of 380,000 t/a of PVC and a corresponding capacity for EDC and VCM intermediates emitted 4,6 t of adsorbable halogenated compounds (AOX) into the effluent. This amounts to an emissions of 13 g AOX] t PVC. Mercury emissions of the same plant, which fully relied on the amalgam process for the manufacture of chlorine and sodium hydroxide were 15 kg per annum.

Vinyl chloride is sometimes proved to be an impurity in ground-water. However, this VCM is a decomposition product of chlorinated hydrocarbons and does not derive from PVC manufacture.

3.5 By-products in the manufacture of vinyl chloride

Integrated oxychlorination: The addition of chlorine to ethylene occurs under very mild conditions. Few by-products are formed [16]. Oxychlorination gives rise to the largest part of by-products. To keep the occurrence of by-products small, the starting products must be as pure as possible. Acetylene in particular gives rise to more highly chlorinated products. By-products also occur during the thermal cracking of EDC. In the literature various information on the composition of the by-products can be found. Table 3.5 is mainly based on indications from Schulze [16]:

TABLE 3.5
By-products of the integrated oxychlorination

Fraction	Quantity
Low boilers	**15 kg/t VCM**
Chloroform	45 %
EDC	17 %
Carbon tetrachloride	9 %
Chlorobutadiene	8 %
1,1-dichlorethane	6 %
Other	15 %
High boilers	**9 kg/t VCM**
1,1,2-trichloroethane	35 %
EDC	8 %
Cis-1,3-dichlorobutene	8 %
Trans-1,3-dichlorobutene	7 %
1,1,2,2-tetrachloroethane	15 %
Tarry residues	**9 kg/t VCM**

VCM manufacture from acetylene and ethylene: The chlorinated hydrocarbon residues are between 10 - 15 kg/t VCM [17]. In the past distillable by-products were treated with chlorine to produce carbon tetrachloride and perchloroethylene. This process is called chlorinolysis. In 1981, 54.9 % of the residues were chlorinolysed, 15 % incinerated with HCl recovery and 30.1 % incinerated on the high seas [16]. This pattern has changed dramatically.

Incineration at sea has completely stopped some years ago. As the demand for perchloroethylene and carbon tetrachloride diminishes, chlorinolysis is also losing significance. Nowadays the major part of chlorinated residues is incinerated with HCl recovery. The recovered HCl is again used in the oxychlorination step.

3.6 PVC dust emissions

In order to comply with the TA-Luft directive, PVC dust emissions must not exceed 50 mg/m^3 with a mass loss greater than 0.5 kg/h. If the mass loss is smaller or equal to 0.5 kg/h, concentrations up to 150 mg/m^3 are allowed. The MAK value for PVC dust is 5 mg/m^3 measured as fine dust [7].

3.7 Ecobalance data (life cycle analysis)

The most recent data on the ecobalance of PVC have been published by BUWAL (Federal Agency for environment, Forest and Landscape, Switzerland). Table 3.6 shows the identified emissions. Note that in Switzerland 80 % of waste is incinerated. The contribution of waste incineration to total emissions is therefore much higher than expected for Germany, where less than 30 % of domestic waste and less than 5 % of total PVC production is incinerated. Nitric oxide and sulphur oxide emissions, attributed to the incineration of PVC, arise from model assumptions, as the contribution of PVC to the nitrogen and sulphur load of waste is negligible.

With respect to production, the data reflect the German situation in 1990 reasonably well. It is most interesting that the major part of emissions into the air does not arise from VCM and PVC production itself. Indirect emissions, caused by mining of petrol and especially by the generation of electricity and heat contribute heavily to these emissions. Often discussed emissions like mercury prove to be of minor importance. This can best be seen by weighing emissions with factors reflecting their ecological impact. Thus critical volumes are derived by dividing the emission by some limit value. Limit values employed by BUWAL [18] can be seen in Table 3.7.

TABLE 3.6
Emissions caused by production and incineration of PVC in g/kg [18]

	Indirect	Process related	Total manufacture.	Waste incineration	Cradle to grave
Atmosphere					
Particles	0.449	0.182	0.631	0.017	0.648
CO	0.816	0.247	1.063	0.34	1.403
HC	9.147	1.372	10.519	0.017	10.536
N_2O	0.634	-	0.634	-	0.634
NO_x	2.242	0.792	3.034	1.332	4.366
SO_2	5.18	0.782	5.962	0.296	6.258
Aldehydes	0.004	0.001	0.005	-	0.005
Cl-organics	-	0.169	0.169	-	0.169
Other organics	0.008	0.5	0.508	-	0.508
NH_3	0.001	-	0.001	-	0.001
HCl	-	-	-	23.216	23.216
Cl_2	-	0.0003	0.0003	-	0.0003
Hg	-	0.00028	0.00028	-	0.00028
Water					
Dissolved solids	12.315	18.529	30.844	-	30.844
AOX	-	0.015	0.015	-	0.015
Other Organics	-	0.524	0.524	-	0.524
Susp. Particles	-	0.062	0.062	-	0.062
COD	0.001	-	0.001	-	0.001
Oil	0.155	0.001	0.156	-	0.156
Phenolics	-	0.005	0.005	-	0.005
Fluorides	0.002	-	0.002	-	0.002
Hg	-	0.00002	0.00002	-	0.00002
Solid Waste	90.4	79.9	170.3	220.8	391.1

TABLE 3.7
Limit values for converting emission data to critical volumes [18]

Atmosphere	mg/m^3	Water	mg/l
Particles	0.07	AOX	0.1
CO	8	Other organics	10
HC	15	Susp.Particles	20
N_2O	0.03	COD	30
NO_x	0.03	Oils	20
SO_2	0.03	Phenolics	0.05
Aldehydes	0.03	Fluorides	10
Cl-organics	0.01	Hg	0.01
Other Organics	0.01		
NH_3	0.5		
HCl	0.1		
Cl_2	0.02		
Hg	0.0007		

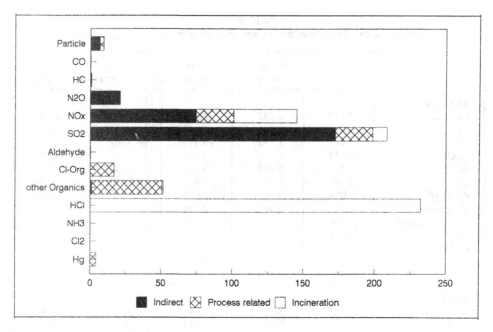

Figure 3.2: Emissions into the air caused by production and incineration of PVC according to BUWAL [18] expressed as critical volumes (1000 m^3/kg PVC)

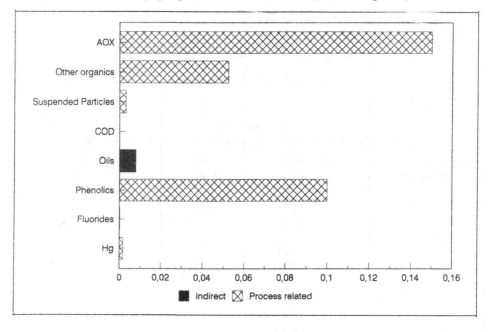

Figure 3.3: Emissions into the effluent caused by production and incineration of PVC according to BUWAL [18] expressed as critical volumes (m^3/kg PVC)

According to BUWAL, NO_x and SO_2 from electricity generation are the most important emissions into the air. If all PVC waste were incinerated, HCl emissions from waste incinerators would be of similar importance. Emissions into the effluent, on the other hand, are predominantly process related. Chlorinated hydrocarbons are the most serious source of water pollution. Some plants use phenolics to stop polymerization. Many companies, however, avoid the use of these compounds on a regular basis. All other emissions into the effluent are of minor importance.

TABLE 3.8

Emissions caused by production and incineration of PVC according to BUWAL [18]
expressed as critical volumes m^3/kg PVC

Critical volumes	Indirect emissions	Process related	Total manufacture.	Waste incineration	Cradle to grave
Atmosphere					
Particles	6,414	2,600	9,014	243	9,257
CO	102	31	133	42	175
HC	609	91	701	1	702
N_2O	21,133	-	21,133	-	21,133
NO_x	74,733	26,400	101,133	44,400	145,533
SO_2	172,666	26,067	198,733	9,86	208,600
Aldehydes	133	33	167	-	166
Cl-organics	-	16,900	16,900	-	16,900
Other Organics	800	50,000	50,800	-	50,800
NH_3	2	2	-	-	2
HCl	-	-	-	232,160	232,160
Cl_2	-	15	15	-	15
Hg	-	4,014	4,014	-	4,014
Total	276,594	126,151	402,746	286,713	689,459
% of total	40 %	18 %	58 %	42 %	100 %
Water					
AOX	-	0.15	0.15	-	0.15
Other organics	-	0.0524	0.0524	-	0.0524
Susp.Particles	-	0.0031	0.0031	-	0.0031
COD	-	-	-	-	-
Oils	0.00775	0.00005	0.0078	-	0.0078
Phenolics	-	0.1	0.1	-	0.1
Fluorides	0.0002	-	0.0002	-	0.0002
Hg	-	0.0017	0.0017	-	0.0017
Total	0.00795	0.30725	0.3152	-	0.3152
% of total	2.5 %	97.5 %	100 %	-	100 %

44

References

1. Rauhut, A., <u>Quecksilberbilanz 1983 - 1985</u>. Landesgewerbeanstalt Bayern, Nürnberg, 1988.

2. Schmittinger, P. et al., Chlorine. In <u>Ullmann's Encyclopedia of Industrial Chemistry</u>, Vol. A6, VCH Verlagsgesellschaft, Weinheim, 1986, pp. 399 - 477.

3. EC-Directive 82/176 of 22.3.1982.

4. TA-Luft, <u>Erste Allgemeine Verwaltungsvorschrift zum Bundesimmissions-schutzgesetz</u> (Technische Anleitung zur Reinhaltung der Luft) of 27.2.1986.

5. Rauhut, A., <u>Quecksilberbilanz 1980 - 1982</u>. Landesgewerbeanstalt Bayern, Nürnberg, 1985.

6. Rauhut, A., <u>Quecksilberbilanz 1977 - 1979</u>. Wasser Luft Betrieb 1982, **5**, 50-51.

7. Maximale Arbeitsplatzkonzentrationen und Biologische Arbeitsstofftoleranzwerte 1990. <u>Mitteilung XXIV der Senatskommission zur Prüfung gesundheitsschädlicher Arbeitsstoffe</u>, VCH Verlagsgesellschaft, Weinheim 1990.

8. Beratergremium für umweltrelevante Altstoffe, Vinylchlorid. <u>BUA Stoffbericht 35</u>, 1989.

9. Callaha, M.A. et al., <u>Water-Related Environmental Fate of 129 Priority Pollutants</u>. Environmental Protection Agency 440/4-79-029, 1979.

10. Radding, S.B. et al., <u>Review of the Environmental Fate of Selected Chemicals</u>. Environmental Protection Agency 560/5-77-003, 1977.

11. Atri, F.R., <u>Chlorierter Kohlenwasserstoffe in der Umwelt II</u>. Gustav Fischer-Verlag, Stuttgart/New York, 1985.

12. Burks, W.M. et al., <u>Wirtschaftlichere Vinylchloridproduktion</u>. Chem. Industrie 1980, **32**, 250.

13. VDI-Richtlinie 2446, <u>Emissionsminderung Vinylchlorid</u>, 1982.

14. Davids, P., Lange, M., <u>Die TA Luft '86 Technischer Kommentar</u>. VDI Verlag 1986.

15. Minott, J.D., <u>A Water Treatment System for a VCM Plant</u>. Chem.Eng. Prog. 1973, **69**, 71.

16. Schulze, J., Weiser, M., <u>Vermeidungs- und Verwertungsmöglichkeiten von Rückständen bei der Herstellung chlororganischer Produkte</u>. Forschungsbericht 103 01 304 UBA-FB 82-128, 1985..

17. Rassaerts, H., Witzel, D., Aliphatische Chlorkohlenwasserstoffe. In <u>Ullmann's Enzyklopädie der technischen Chemie</u> 4.Ed. Vol. 9, VCH Verlagsgesellschaft, Weinheim, 1975, pp. 404 - 498.

18. Habersatter, K., Widmer, F., <u>Oekobilanz von Packstoffen. Stand 1990</u>. Bundesamt für Umwelt, Wald und Landschaft (BUWAL), Schriftenreihe Umwelt Vol. 132, Bern, 1991.

4. Processing of polyvinyl chloride

4.1 Processing methods

PVC is converted by all techniques applied for thermoplastic materials. By far the most important process is extrusion, followed by calendering. Paste processing occupies the third place among the processing methods. In paste processing sufficient plasticizer is added to the resin to enable it to be processed as a paste. This processing mode is limited to PVC. In Germany injection moulding, blow moulding and compression moulding are less important processing techniques for PVC. Table 4.2 shows the importance of each of the processes.

The energy requirement for the processing of PVC is around 5 to 10 % of the energy expended in its manufacture. Table 4.1 displays the energy consumption required for processing PVC. To enable the latter to be compared with the energy consumption for the manufacture of polyvinyl chloride, the energy equivalence value is given in [1]. For processing, this corresponds to the primary energy requirement needed to make available the necessary electrical energy (1 kWh current corresponds to 10.2 MJ primary energy), while the energy equivalence value for the manufacture of PVC embraces the total energy content of the raw materials and the energy requirement for electrolysis, process heat, pumps, etc.

TABLE 4.1

Energy requirement for plastics processing expressed as energy equivalent [1]

Processing mode	Energy consumption (GJ/t)		
Film extrusion	3	-	6
Film calendering		6	
Pipe extrusion	3	-	5
Blow moulding	5	-	15
Injection moulding	5	-	15
Manufacture of PVC		53	

TABLE 4.2

Distribution of PVC processing methods (FRG 1980) according to Finkmann [2]

	Extrusion %	Extrusion kt	Injection %	Injection kt	Calendering %	Calendering kt	Blow %	Blow kt	Paste %	Paste kt	Compr. moulding %	Compr. moulding kt
Flexible PVC												
Wire insulation	100.0	90.0	-	-	-	-	-	-	-	-	-	-
Flexible film	5.9	5.0	-	-	94.1	80.0	-	-	-	-	-	-
Floor coverings	-	-	-	-	33.3	20.0	-	-	66.7	40.0	-	-
Profiles, hoses	100.0	40.0	-	-	-	-	-	-	-	-	-	-
Coatings, pastes	-	-	-	-	-	-	-	-	100.0	70.0	-	-
Injection mouldings	-	-	100.0	10.0	-	-	-	-	-	-	-	-
Total PVC-P	38.0	135.0	2.8	10.0	28.2	100.0	-	-	31.0	110.0	-	-
Rigid PVC												
Films, slabs, sheets	12.1	20.0	-	-	87.9	145.0	-	-	-	-	-	-
Window profiles	100.0	130.0	-	-	-	-	-	-	-	-	-	-
Other profiles	100.0	130.0	-	-	-	-	-	-	-	-	-	-
Pipes, guttering	94.0	235.0	6.0	15.0	-	-	-	-	-	-	-	-
Blow mouldings	-	-	-	-	-	-	100.0	25.0	-	-	-	-
Injection mouldings	-	-	100.0	5.0	-	-	-	-	-	-	-	-
Records	-	-	-	-	-	-	-	-	-	-	100.0	25.0
Others	-	-	-	-	-	-	-	-	-	-	-	-
Total PVC-U	70.5	515.0	2.7	20.0	19.9	145.0	3.4	25.0	-	-	3.4	25.0
Total	59.9	650.0	2.8	30.0	22.5	245.0	2.3	25.0	10.1	110.0	2.3	25.0

4.2 Composition of PVC compounds

4.2.1 PVC types

The range of variables in formulating PVC begins with the selection of the right PVC type. Which type is used will depend on the processing method and on current price. Suspension (S)- and mass (M)-PVC are interchangeable in many fields of application [2]. When a high degree of transparency is demanded, M-PVC is preferred because of its purity. S-PVC is dominant in processing plasticized material, notably cable and wire sheathings. Micro-S-PVC is used almost exclusively for paste processing. Table 4.3 shows which type is used for which application.

TABLE 4.3
Application of PVC types

PVC type	S-PVC	M-PVC	Micro-S	E-PVC
Flexible-PVC				
Wire insulation	++	+	-	-
Flexible film	++	+	-	+
Floor coverings	++	+	-	+
Profiles, hoses	++	+	-	-
Coatings, pastes	+	+	++	++
Injection mouldings	++	+	-	-
Rigid PVC				
Films, slabs, sheets	++	+	-	+
Window profiles	++	-	-	-
Other profiles	++	++	-	++
Pipes, guttering	++	++	-	+
Blow mouldings	+	++	-	-
Injection mouldings	++	++	-	-
Records	++	+	-	+

The pure polymer is almost never processed. Additives facilitate smooth processing and determine the properties of the product. Rigid PVC (PVC-U) contains less additives than plasticized PVC (PVC-P), although a chlorine-free copolymer is added in concentrations of 5 to 8 % to PVC-U in order to enhance the impact strength. Flexible PVC may contain 30 to 60 % plasticizer so that, in an extreme case, the PVC product contains only 30 % of polyvinyl chloride.

4.2.2 Additives

The compatibility of PVC with a large number of additives allows for an extremely broad product range from pastes for car underbody protection to window profiles. Additives determine the mechanical properties, light and weathering fastness, colour, electrical resistance and other properties.

TABLE 4.4
Effect of PVC additives on product properties [3]

Auxiliary	Affected property of the PVC article
Stabilizer	Prevents decomposition during processing, imparts light and weathering resistance
Colorant	Colour, weathering resistance
Plasticizer	Mechanical properties, burning behaviour
Impact modifier	Impact strength and other mechanical properties
Lubricants	Rheology of the PVC melt, also affects transparency, gloss, surface finish and printability
Fillers	Mechanical properties
Flame retardants	Burning behaviour
Antistatic agents	Electrical properties
Blowing agents	Processing to expanded products

The effect of these constituents is not additive. Desirable as well as undesirable synergisms may occur. Matching of additives with the stabilizer hence is an absolute necessity. Furthermore, different processing methods make different demands on the stability and rheology of the PVC melt.

Stabilizers

The stabilizers used in the Federal Republic are lead compounds, Ba/Cd soaps, Ba/Zn or Ca/Zn soaps and tin compounds. Stabilizers based on antimony or those of a purely organic nature play a secondary role. However, the stabilizers are frequently combined with organic co-stabilizers such as epoxides, polyols, phosphites and antioxidants in order to optimize weathering behaviour. Every stabilizer has typical uses even though the fields of application overlap. For example, the electrical properties of lead compounds are so outstanding, that presently all cable compounds are stabilized with lead. Table 4.5 assigns the customary stabilization to the product groups.

TABLE 4.5
Areas of application of stabilizers [3]

Stabilizer	Ba/Cd	Ba/Zn	Ca/Zn	Pb	Sn
Flexible PVC					
Wire insulation	-	-	-	++	-
Flexible films	(+)	++	+	-	-
Floor coverings	-	++	-	-	-
Profiles, hoses	(+)	++	++	-	-
Coatings, pastes	(+)	++	-	-	-
Injection mouldings	(+)	+	-	+	-
Rigid PVC					
Films, slabs, sheets	+	-	-	+	++
Window sections	++	-	-	++	-
Other profiles	+	-	-	++	+
Pipes, guttering	-	-	-	++	-
Blow mouldings	-	-	++	-	++
Injection mouldings	-	-	-	++	-
Records	-	-	-	-	++

In terms of quantity, the **lead stabilizers** are of outstanding importance (cf. Table 4.6). They are favourably priced and can be employed in higher doses. By suitably combining the lead compounds it is possible to cover a broad spectrum of demands. Their high heavy metal content counts as a disadvantage. They cannot be used for pure white or transparent products.

TABLE 4.6
Quantitative use of PVC stabilizers (FRG 1985) [4]

Stabilizer type	Consumption (t)	Metal content (t)	
Pb stabilizers	13,410	8,448	Pb
Sn stabilizers	3,328	478	Sn
Ba/Cd stabilizers	2,630	277	Cd
Ba/Zn stabilizers	3,095	310	Ba/Zn
Ca/Zn stabilizers	80	6	Ca/Zn
Total	**22,553**	**9,519**	

The **tin stabilizers** are differentiated according to those containing sulphur and those which are free from sulphur. Thio-tin stabilizers accommodate the highest processing temperatures and the manufacture of highly transparent products. Unfortunately these products have little light-fastness. By contrast, sulphur-free tin stabilizers impart optimum protection to the polymer against light induced decomposition. But they stabilize less satisfactorily against

50

thermal damage and give rise to more problems in processing. Selected tin stabilizers are permitted for use in PVC products for food contact. During processing tin stabilizers may split off mercaptides, disulphides or maleic acid, causing pollution of the air at the working place and in the environment of the factory.

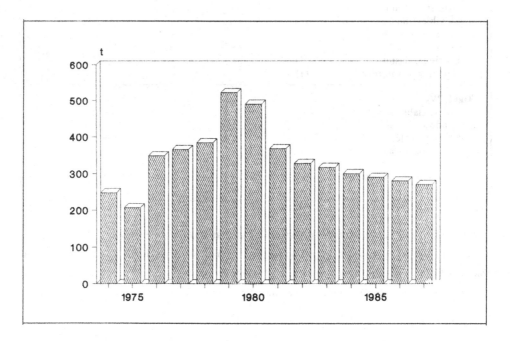

Figure 4.1:   Use of cadmium for PVC stabilization [3]

Barium/Cadmium stabilizers are used for durable exterior products. Substantial safety in processing together with good light stabilization are the reasons for their importance in the manufacture of window sections, sidings and of roofing sheets. Figure 4.1 shows the development of the use of cadmium in PVC stabilizers. The development of cadmium substitutes has reached a stage where a noticeable decline in the consumption of cadmium is to be expected from 1992 or 1993 onwards [3].

The stabilization mechanism of **Barium/Zinc** and **Calcium/Zinc** stabilizers is similar to that of Ba/Cd stabilizers. However, reaction with hydrogen chloride forms zinc chloride which has a stronger destabilizing effect than cadmium chloride. For this reason Ca/Zn stabilizers originally were developed only for use in products like toys or food packaging. By now this system has been mastered so well that improved Ca/Zn stabilizers are recognized as the most interesting alternatives to Ba/Cd stabilizers for PVC-U based exterior products.

TABLE 4.7
Use of Cd stabilizers in various fields of application (FRG 1985) [4]

Product	t stabilizer	t Cd
Window profiles	1800	210
Other exterior profiles	390	47
Roof coverings	400	18
Plastisols	10	<1
Extrusion of plasticized compounds	120	4
Total	**2720**	**280**

Pigments

Polyvinyl chloride is processed at temperatures around 180 °C so that only low demands are placed on the temperature fastness of the pigments. As a result the use of high temperature resistant pigments such as cadmium sulphoselenides is unnecessary. Cadmium pigments hence are not used for the colouring of PVC. Resistance to migration is of great importance for pigments used in plasticized PVC. Pigments which are partially soluble in the polymer or plasticizer would bleed or bloom out. All inorganic pigments are fast to migration, but there are also many organic pigments which meet this requirement. The demands with respect to light and weathering fastness depend on the application. The demands placed on the chemical resistance will depend on the formulation and on the environmental effects. Table 4.8 gives an overview of the most important inorganic pigments for colouring of PVC.

TABLE 4.8
Inorganic pigments for colouring of PVC

Pigment	Composition
Chrome yellow	$Pb(Cr,S)O_4$
Nickel-titanium yellow	$(Ti,Sb,Ni)O_2$
Chromium-titanium yellow	$(Ti,Sb,Cr)O_2$
Iron oxide yellow	$FeOOH$
Molybdate red	$Pb(Cr,Mo,S)O_4$
Iron oxide red	Fe_2O_3
Ultramarine blue	$Na_8Al_6Si_6O_{24}S_{(2-4)}$
Chrome-iron brown	$(Fe,Cr)_2O_3$
Titanium dioxide (white)	TiO_2
Carbon black	

Phthalocyanines are the most important organic pigments for colouring PVC. They dominate the blue and green colour range. Monoazo pigments contain an azo group as chromophore.

They exist in the colour shades yellow, orange, red and brown. Table 4.9 gives an overview
of the most important organic pigment classes:

TABLE 4.9
Important organic pigments for colouring PVC

Class of pigments	Shades
Phthalocyanines	blue, green
Monoazo pigments	yellow, orange, red, brown
Disazo pigments	yellow, red
Disazo condensation pigments	yellow, red, brown
Azomethine pigments	yellow, orange
Naphthalene and perylene pigments	orange, red
Anthraquinone pigments	red, yellow, blue
Quinacridone pigments	red, red-violet
Dioxazine	violet

Plasticizers

About one third of polyvinyl chloride is processed with the aid of plasticizers. The most
important applications are cable compounds, flexible films, floor coverings and plastisols. As
shown in table 4.10, 214 kt phthalic acid based plasticizers were used in the Federal
Republic in 1987. By far the most important compound is dioctyl phthalate, or more exactly
di-(2-ethylhexyl) phthalate (DOP or DEHP). Esters of phosphoric acid, of aliphatic
dicarboxylic acids (adipic, sebacic, azelaic) and some speciality products are also used. The
manufacture of flexible PVC consumes 77 % of the plasticizer production, the remainder is
used for plasticizing cellulose derivatives, paints and adhesives.

TABLE 4.10
Production, import and export of plasticizers (FRG 1987)

Plasticizer	Production	Import	Export
Dibutyl phthalate	22,279	15,506	14,736
Dioctyl phthalate	251,439	18,962	182,449
Diisooctyl, diisononyl, Diisododecyl phthalate	48,721	44,216	32,303
Other phthalates	58,549	-	15,650
Aliphatic dicarboxylic acids	27,504		

4.3 Emissions in processing

Processing of PVC can give rise to emissions, which usually originate from the additives. PVC converters who are located in the neighbourhood of domestic housing have to meet particularly stringent conditions. The exhaust air may be purified by the following procedures [5]:

- physical separation of aerosols
- adsorption
- catalytic post-combustion or
- thermal post-combustion

Possible emissions will be briefly considered here:

Vinyl chloride: Using state of the art technology, plastics manufacturers supply a polymer with a low residual monomer content. This amounts to 2 ppm for suspension PVC, 5 ppm for emulsion PVC and 2 ppm for mass PVC [6]. Individual types can have a distinctly lower monomer content. On processing, the monomer content falls to 0.1 ppm [6]. As no vinyl chloride is formed on the decomposition of PVC, the upper limit for the sum of VCM emissions from processing, use and destruction of PVC in the Federal Republic is 5 t.

Hydrogen chloride is only formed in the uncontrolled decomposition of PVC. Such decomposition is carefully avoided, because HCl causes rapid corrosion of expensive processing equipment. HCl emissions during processing are not likely to occur, unless there is a breakdown in production.

Plasticizers: Plasticizers are chosen among substances having an extremely low vapour pressure. Under processing conditions, however, plasticizer vapours and decomposition products occur. The working place concentration of di-(2-ethylhexyl) phthalate (DEHP) must not exceed 10 mg/m^3 [7]. TA-Luft 86 gives DEHP a Class II rating, i.e. an exhaust gas concentration of 0.10 g/m^3 must not be exceeded [8]. In the production of spread coated products, the amount of process air purged to the environment is typically 15 - 25 m^3/kg plasticizer consumed. If the air is filtered according to state of the art technology, the concentration of plasticizer passing to environment is at most 20 mg/m^3 [9].

Stabilizers: Due to the toxic nature of cadmium, barium/cadmium stabilizers are overwhelmingly used in a dust-free form or according to the principle of 'lost packaging'. When maleinate containing tin stabilizers are used, the release of maleic acid can lead to irritations.

Thio-tin stabilizers can create an odour nuisance because they split off mercaptides and disulphides.

Decomposition products of **blowing agents:** PVC foams are blown with agents which form the foam either by vaporization (physical blowing agents) or by decomposition (chemical blowing agents). The most important chemical blowing agent (about 95 %) is azodicarbonamide which decomposes to give nitrogen, carbon monoxide, carbon dioxide and solid, sublimable compounds (Urazole, cyanuric acid, hydrazodicarbonamide, cyammelide) [10].

Solvents are used in PVC processing for the formulation of organisols and plastisols (about 2.5 kt/year) and in some additives (0.5 kt/year). This accounts for 0.3 % of solvent consumption in the Federal Republic. The solvents mainly involved are aliphatic hydrocarbons [11]. **Pigment dusts** can be avoided by using the pigments as pastes or in another dust-free form.

4.4 Examples of PVC products

4.4.1 Window profiles

Next to pipes, window profiles are the most important application of PVC in the building industry. Notwithstanding a general stagnation in the building trade, the PVC window was able to realize vigorous growth as it replaced worn out wooden windows in the renovation of old buildings. At the present time about 40 % of all windows are fabricated with PVC sections, a further 40 % from wood and 20 % from aluminium. Additionally, there is a small amount of window frames from polyurethane foam. As recently as 1970 the market share of PVC windows was only 3 to 4 % [12]. Table 4.11 shows the development of the quantity produced since 1982.

TABLE 4.11

Production of window profiles in the FRG according to Federal Statistical Office data [13]

Year	t	1000 m	Year	t	1000 m
1982		115,910	1987	152,155	168,353
1983	129,946	133,558	1988	188,447	188,199
1984	125,211	133,485	1989	201,787	211,087
1985	131,017	137,108	1990	211,730	233,563
1986	136,344	148,012			
Goods production number: 5831 13 [13]					

The Federal Republic exports far more window profiles than it imports. The association of plastics converters (GKV) therefore estimates that the consumption of window profiles within the country is distinctly lower than the production [14].

TABLE 4.12
Consumption of PVC window profiles in the FRG according to GKV data [14] (tonnes)

Year	Quantity	Year	Quantity	Year	Quantity
1963	185	1972	24,000	1981	115,000
1964	397	1973	31,000	1982	108,000
1965	619	1974	33,000	1983	123,000
1966	1,008	1975	41,000	1984	110,000
1967	1,541	1976	50,000	1985	90,000
1968	2,169	1977	60,000	1986	92,000
1969	3,188	1978	85,000	1987	96,000
1970	7,000	1979	115,000		
1971	14,000	1980	130,000		

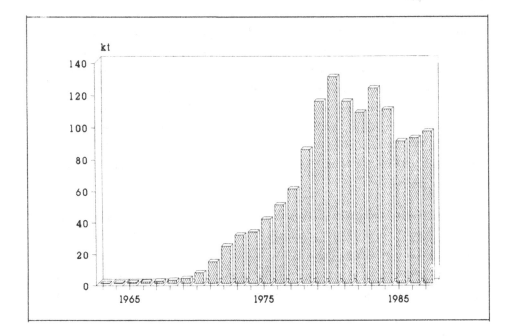

Figure 4.2:   Consumption of PVC window profiles in the FRG in tonnes according to information given by GKV [14]

In Germany, window profiles are manufactured by about 60 firms. Some 1000 companies fabricate windows from these sections. The profiles are worked by sawing, milling and drilling in a manner similar to that of wood. Corners, however, are joined by welding. By automating these processes the average production time has been reduced from 2.5 to 1 hour. As a result windows from wood and PVC have become approximately equal in price [12]. The UK and Austria are the only countries where the PVC window plays as dominant a role as in the FRG. The market in France and the Scandinavian countries is expected to grow substantially in the near future. The Federal Republic imports small quantities of window sections from Austria and Italy.

Formulation

Base material: The material used without exception is suspension PVC with a K value of 64 to 70. The polymer is modified with 6 to 8 % acrylate. This greatly improves toughness. Previously graft polymers with ethylene-vinyl acetate or polymer blends with chlorinated polyethylene were employed.

Stabilizer: Stabilization is extremely important because window profiles are expected to last for a long time resisting sun and weather. Also the complex extrusion geometry is very demanding. This places window profiles into the domain of the barium/cadmium stabilizers, with 75 % of all cadmium stabilizers being used in the manufacture of window profiles [3]. In the last few years there has been a transition from pure Ba/Cd stabilization to combinations with lead. In the acrylate modified PVC base materials which are commonly used today, these Ba/Cd/Pb stabilizers offer the optimum in process safety. Lead stabilization has grown in importance over the last years. Tin stabilization is used only in the manufacture of coloured window profiles.

In the near future improved Ca/Zn-stabilizers are likely to replace cadmium and lead containing formulations for the stabilization of window profiles. Originally, the Ca/Zn combination was only used in products like toys or food packaging. By using new types of organic co-stabilizers the effectiveness of Ca/Zn stabilizers could be increased to such an extent that they equal cadmium stabilized products in accelerated weathering tests. However, no experience has yet been gained with built-in windows. Once these stabilizers have proved themselves in durability tests it is likely that most manufacturers will change to them in their formulations. However, Ca/Zn stabilization will cost roughly 50 % more than cadmium or lead based stabilization [3].

TABLE 4.13
Formulation of PVC compounds for window profiles (acc. to Polte [15] and Tötsch [3]),
additions per 100 parts acrylate modified PVC

Constituents	Ba/Cd	Ba/Cd/Pb	Pb	Sn
Ba/Cd stabilizer	2.0 - 3.0	1.0 - 1.5	-	-
Lead phosphite	-	1.5 - 3.0	3.5 - 4.0	-
Tin maleate	-	-	-	2.5 - 3.0
Organic phosphite	0.3 - 0.5	-	-	-
Epoxidized soya oil	0.5 - 1.5	-	-	-
Internal lubricant	0.8 - 1.2	0.8 - 1.2	0.8 - 1.8	0.4 - 0.8
External lubricant	0.2 - 0.4	0.2 - 0.4	0.2 - 0.6	0.6 - 1.2
Chalk	3.0 - 5.0	3.0 - 5.0	2.0 - 4.0	2.0 - 4.0
Titanium dioxide	2.0 - 4.0	2.0 - 4.0	2.0 - 4.0	2.0 - 4.0
Process aid (PMMA)	0.8 - 1.2	0.8 - 1.2	1.0 - 2.5	1.0 - 2.5
Use frequency	35 %	50 %	15 %	low

4.4.2 Floor coverings from polyvinyl chloride

Floor coverings from PVC are a very inhomogeneous product group. Only calendered floor
coverings, cushioned vinyls and floor coverings made by spread coating still enjoy large
scale manufacture. There is a shift in emphasis away from cheap spread coated coverings to
higher value products [16]. Rigid vinyl tiles are hardly produced any more The oldest form
of PVC floor covering, the vinyl asbestos tile, is no longer made because of the carcinogenic
potential of asbestos. Also around 4000 t PVC are used for carpet backing. In this
application the PVC serves to anchor the pile and improves the layability of the carpet rolls
[12].

TABLE 4.14
Use of PVC in floor coverings in tonnes (FRG 1987)

Nature of floor covering	Covering	PVC content
Calendered coverings	55,000	18,000
Cushioned vinyls	42,000	21,000
Spread coated coverings	8,000	5,000
Carpet backing	-	4,000
Total	105,000	48,000

The fall in demand for PVC floor coverings is attributed to the competition from carpeting.
Carpeting is predominant in living-rooms, bedrooms, dining-rooms, children's rooms and
entrance halls, while PVC floor coverings have been able to maintain their position in

58

bathrooms and kitchens [12]. The total manufacture of PVC floor covering in 1987 was around 105 kt. This, however, corresponds to only about 48 kt pure polyvinyl chloride. The difference is explained by the high proportion of carrier material, filler and plasticizer.

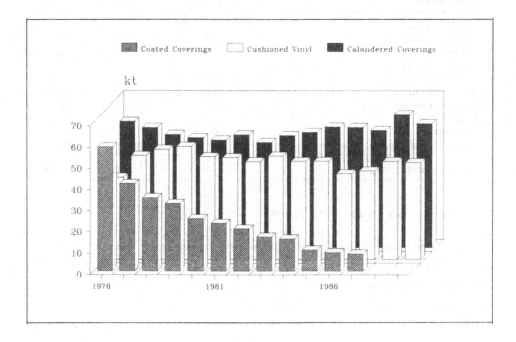

Figure 4.3:   Development of plastics floor covering production in the FRG [13]

Calendered floor coverings

These flexible floor coverings are supplied in rolls up to 4 m wide or in sheets [16]. The thickness varies between 1.5 and 3 mm. These floor coverings are mainly for heavy duty use. Most floor coverings have a multi-layer structure - up to 4 layers are usual. In homogeneous coverings all layers are similarly formulated. The covering contains little filler and is highly abrasion resistant. As the whole floor covering can act as a wear layer, it is suitable for heavy duty service in public buildings. In heterogeneous coverings only the wear layer contains little filler and is therefore abrasion resistant. For economic reasons the lower layers are highly filled. Heterogeneous coverings are predominantly used in normal living areas.

TABLE 4.15:   Formulation of a calendered floor covering

Constituent	Parts by weight
S-PVC, K value = 70	100
Ba/Zn stabilizer	3 - 5
Dioctyl phthalate	40 - 60
Lubricant	0.5
Chalk	100 - 200

Cushioned vinyls (CV)

It is mainly pattern and colour which sell floor coverings. The price difference between high value CV coverings and carpeting is no longer important. On average the customer buys only 13 to 14 m^2 CV covering while the average costumer buys some 24 m^2. This is explained by the smaller area occupied by kitchen, bathroom or entrance hall [17]. About half the CV floor coverings are laid by customers themselves (DIY).

The heart of a CV floor covering is a backing material bedded into the PVC [17]. These days, a glass mat is used almost exclusively as the backing material. This glass mat is 0.4 mm thick, with about 600 g plastisol being used per square metre. This layer is slightly gelled before being coated with a second layer of PVC. This is the foam layer which contains a chemical blowing agent, usually azodicarbonamide. On heating azodicarbonamide decomposes with the evolution of gas. Activators - also referred to as kickers - accelerate the decomposition of the blowing agent. Before expansion the layer thickness is about 0.4 mm. This layer is slightly gelled, i.e. it is not heated to such an extent that the kicker reaction is activated. The layer is now printed. One part of the ink contains an inhibitor. Inhibitors impair the action of kickers, i.e. the plastisol does not expand at those places where this ink is printed. A clear top coat is applied over the printed surface. This coat is about 0.3 mm thick and is left unfilled to ensure the greatest possible wear resistance. The whole composite is now heated for 2 minutes at 200 °C. In those places where the foam layer contains no inhibitor the azodicarbonamide decomposes causing the PVC to expand. The result is a structured surface. To further enhance sound and heat insulation an expanded layer is frequently applied under the glass mat. This foam is mostly made mechanically by mixing in air.

TABLE 4.16
Formulation and operating condition for the manufacture of a CV floor covering according to Skura [18]

Functional layer	Formula	Parts	Processing conditions
Glass mat backing	Paste PVC	100	Application: 600 g/m^2
	Dioctyl phthalate	75	slightly gelled over a
	Filler (CaCO$_3$)	130	gelling drum (140 °C,
	Thio tin stabilizer	0.4	contact time 20 sec)
Expanded layer	Paste PVC	70	Application: 0.4 mm,
	Extender PVC	30	slightly gelled at
	Dioctyl phthalate	40	140 °C for 2 min
	Butyl benzyl phthalate	10	
	Epoxidized soya oil	3	
	Filler (fine, light chalk)	25	
	TiO$_2$	5	
	Azodicarbonamide	2	
	ZnO	0.9	
Inhibitor containing printing ink	Binder (40 % PU dispersion in water)	51	The printing ink is combined in a ratio of 90:10
	Water	10	with inhibitor and is
	Ethanol	19	printed on directly
	Isopropanol	14	
	Pigment	6	
Transparent top coat	Micro-S PVC	70	Application: 0.3 mm,
	Extender PVC	30	expansion of the whole
	Dioctyl phthalate	30	composite: 2 min at
	Butyl benzyl phthalate	20	200 °C
	Epoxidized soya oil	3	
	Thio tin stabilizer	2	

Spread coated floor coverings

These are low priced floorings in which PVC paste is coated onto a textile carrier. The fully fused PVC forms the wear coat. These coverings too can have a structured surface which is achieved by means of embossing and printing. The flexible carrier enables these coverings to have a better footfall sound and thermal insulation. However, they are less suitable for rooms where water spills may occur [12]. The carrier used is jute or polyester felt [16]. A PVC base coat is applied to this backing. In the doctor knife process pregelling is followed by application of a 0.1 to 0.3 mm thick wear coat which is then fully fused. This coat is embossed and a plastisol of the same composition but different colouring is then coated into the indentations by doctor knife. After final fusion the surface can still be structured by means of an embossing roll. The optical effect can be created by printing processes involving printing on the fully fused base coat. After drying of the print, a 0.1 - 0.3 mm thick transparent coat is applied and the web is fully fused.

TABLE 4.17
Formulation for the wear coat of spread coated floor coverings

Constituent	Parts by weight		
Micro-S PVC, K value = 70		100	
Ba/Zn stabilizer	1	-	1.5
Plasticizer blend	40	-	60
Epoxy plasticizers	2	-	4

Vinyl asbestos tiles

This floor covering was manufactured by calendering and then cut into tiles. This type of floor covering is not produced any more because of the low footfall sound and thermal insulation and, most important, because of the dangers attached to asbestos [12].

References

1. Kindler, H., Nikles, A., Energieaufwand zur Herstellung von Werkstoffen - Berechnungsgrundsätze und Energieäquivalenzwerte von Kunststoffen. Kunststoffe, 1980, **70**, 802 - 807.

2. Finkmann, H.-U., Einführung in die Verarbeitung von PVC. In Kunststoff-handbuch. Vol. 2 Polyvinylchlorid, Carl Hanser Verlag, München Wien 1986, pp. 870 - 876.

3. Böhm, E., Tötsch, W., Cadmiumsubstitution - Stand und Perspektiven. Verlag TÜV - Rheinland, Köln, 1989.

4. Data from the Association of plastics manufacturing industry, VKE.

5. Feikes, L., Emissionen beim Betrieb von Beschichtungs-, Kalander- und Verede-lungsanlagen - Meßtechnische Probleme. Kunststoffe 1981, **71**, 86 - 90.

6. Beratergremium für umweltrelevante Altstoffe, Vinylchlorid. BUA-Stoffbericht 25 VCH Verlagsgesellschaft, Weinheim, 1989.

7. Maximale Arbeitsplatzkonzentrationen und Biologische Arbeitsstofftoleranzwerte 1988, Mitteilung XXIV der Senatskommission zur Prüfung gesundheitsschädlicher Arbeitsstoffe, VCH Verlagsgesellschaft, Weinheim, 1988.

8. Erste Allgemeine Verwaltungsvorschrift zum Bundesimmissionsschutzgesetz (Technische Anleitung zur Reinhaltung der Luft) of 27.2.1986.

9. Menzel, B., private commnication.

10. Hurnik, H., Chemische Treibmittel. In Kunststoffadditive. Carl Hanser Verlag, München Wien, 1983, pp. 635 - 656.

11. Federal Environment Office, private communication.

12. v.Bassewitz, A. et al., Anwendung von PVC. In Kunststoffhandbuch, Vol. 2 Polyvinylchlorid, Carl Hanser Verlag, München Wien 1986, pp. 1187 - 1236.

13. Produktion im produzierenden Gewerbe des In- und Auslands. Fachserie 4, Reihe 3.1, Published by, Statistisches Bundesamt, Wiesbaden, 1967 - 1987.

14. Data from the Trade Association for Building, Furniture and Industry - semi-finished goods from plastics in GKV.

15. Polte, A., Schreiter, H.U., Herrmann, H.-D., Extrudieren von Rohren, Profilen, Schläuchen und Ummantelungen. In Kunststoffhandbuch. Polyvinylchlorid, Vol. 2, Carl Hanser Verlag, München Wien, 1986, 905 - 950.

16. Kunststoffböden 1979. Kunststoffe im Bau 1979, **14**, 158 - 160.

17. Cushioned Vinyl-Beläge, Tendenzen. Boden Wand Decke 1986, 1, 5 - 7.

18. Skura, J., Inhibieren von Treibmitteln für chemische PVC-Schäume. Kunststoffe 1986, **76**, 501 - 503.

5. Polyvinyl chloride in waste

5.1 Occurrence of PVC residues

In 1987, some 1,320 kt PVC were manufactured in the Federal Republic [1]. Allowing for an export trade surplus of PVC moulding compounds (260 kt), PVC semi-finished goods (160 kt) and PVC in finished products or as packaging material (100 kt), approximately 800 kt pure PVC remained in the Federal Republic. All the figures are calculated with reference to the pure polymer. The export trade surplus with respect to moulding compounds and semi-finished goods can be obtained from the export trade statistics [2], while the figure for the export trade balance of PVC in finished products is estimated. The 800 kt PVC remaining on the home market corresponded to a decidedly greater quantity of PVC products because significant amounts of additives have been added to the polymer during processing.

Calculated on the basis of pure PVC, manufacturing industry generated about 150 kt PVC waste in 1987, of which 60 kt were channelled to the recycling trade. A further 70 kt were disposed of through the municipal waste disposal system. The remaining 20 kt commercial waste were deposited on company owned waste disposal sites or removed to special waste sites. In 1987 around 85 kt PVC found its way into domestic waste. As shown in Table 5.3, the majority of this PVC is made up of packaging material [4]. The PVC content of bulky refuse can only be estimated at around 15 kt. In addition, there are other waste streams, e.g. with automobile scraps (approx. 40 kt) or as part of demolition waste. The discrepancy between the amount found in waste and PVC consumption arises from the durability of PVC products. There are currently about 7,000 kt PVC in use. The largest part of polyvinyl chloride is converted into building products which are more than 10 years in service. These PVC products will therefore begin to appear as waste only at some future time when buildings are pulled down or converted. Given an approximately constant use of PVC in the building trade, there will be an annual quantity of about 500 kt PVC waste from demolition waste 10 to 15 years from now. Figure 5.1 gives an overview from which, in the interest of clarity, small volume waste streams, such as PVC in automobile scrap, have been omitted.

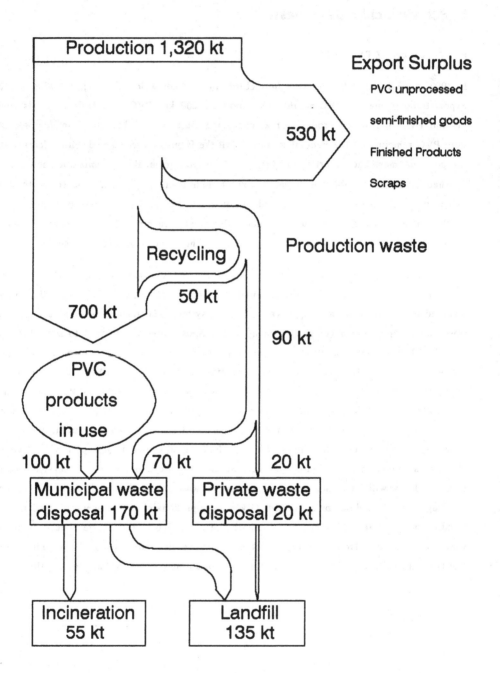

Figure 5.1: Flow of Polyvinyl chloride in 1987

5.2 Waste occurring in manufacturing industry

In the industrial processing of PVC, waste occurs directly in the form of off cuts, sprues, residues from pressing or faulty batches and after the use of PVC products. In 1987 the quantity of waste is likely to have been at least 150 kt (referred to the pure polymer). Industrial PVC residues are most suitable for collection and reuse: They occur concentrated at several points and are of consistently high quality.

Information on where and in which quantity PVC residues occur in industrial manufacture can be obtained from the official statistics on waste disposal [3]. This survey has been carried out only since 1980 and still has some failings. Thus from 1980 to 1984 the quantity of waste appears to have grown more strongly than the PVC production. This growth is probably not real: The newer figures are simply more accurate. However, there are still some branches of industry processing large quantities of PVC which hardly ever report any waste. This is the reason why the quantity for residues of 142 kt (inclusive additives) reported for 1984 will also be too low. Despite these failings there are interesting trends to be gathered from the statistics on waste disposal (Table 5.1):

TABLE 5.1
Occurrence and retention of PVC residues in manufacturing industry according to Federal
Statistical Office data [3] (tonnes)

Year	Quantity of waste	External disposal	Company owned landfill	incinerat.	Recycling (external)
1980	91,062	58,560	7,687	28	24,795
1982	107,407	58,592	13,367	4	35,444
1984	142,384	78,144	12,297	15	51,927
1987	158,652	81,327	12,805	58	59,492

* PVC including additives. Figures also include the small quantity of fluoropolymer waste

In 1984, 36 % of the quantity of waste was passed on to the raw materials trade or to firms carrying out further processing work. In absolute figures double the quantity of PVC residues was reused compared to 1980. These residues were not exclusively processed within the country. According to [2], there is an export trade surplus in PVC residues of about 10 kt. Incidentally, recycling within the company is not recorded by the statistics on waste disposal. Unused quantities of residues were for the most part picked up by the refuse collection service (23 %) or brought to domestic waste disposal sites of the relevant authority (29 %). A lot less PVC waste was absorbed by factory owned landfills (8.6 %) or special waste sites (2.6 %). The amount of PVC residues disposed of by factory internal incineration is insignificant.

It was hardly surprising that most plastics waste occurred in the manufacture of plastics goods. A total of 536 manufacturers reported 83 kt waste of which 47 % was reused. If factory internal recycling would also be recorded, the recycling proportion would be decidedly greater. The chemical industry generated 20 kt PVC residue. This originated in part, but not exclusively, from plastics production. Thus the largest part of the identified waste already occurs before use of the plastics goods. Larger amounts of residues also occurred in the electrical industry, road vehicle construction and in paper processing. The direct building trade, which is one of the most important consumers of PVC products, reported a strikingly low quantity of 97 t PVC residues. Just the off-cuts arising from fitting pipes, guttering and cable ducts would have had to result in a markedly greater quantity.

TABLE 5.2
PVC waste* in the manufacturing industry (FRG 1984) according to Federal Statistical Office data [3]

Industry field	Quantity of waste (t)	Recycling (t)	Proportion (%)
Plastics converters	83,827	39,699	47
Chemical industry	20,195	427	2
Electrical industry	6,897	3,856	56
Road vehicle construction	6,813	576	8
Paper processing	5,929	2,056	35
Leather working	3,974	524	13
Textile industry	3,033	489	16
Printing	2,156	1,466	68
Rubber processing	1,855	420	23
Wood working	1,519	302	20
Musical instruments, toys etc.	544	284	52
Building extension trade	521	395	76
Precision engineering, optics	173	19	11
Steel, light metal construction	155	66	43
Glass	144	143	99
Direct building trade	97	64	66
Manufacturing industry	**142,384**	**51,927**	**36**

*PVC including additives. Figures also include the small quantity of fluoropolymers

5.3 Scrap occurring in domestic refuse

Despite its substantial economic importance, relatively little PVC finds its way into domestic refuse. In 1985, Brahms et al. found 111.3 kt PVC in domestic refuse, a figure which corresponds to 0.79 % of the total accumulation of domestic refuse. 111.3 kt processed PVC correspond to 85 kt polymer which amount to 6.5 % of annual production. The reason for this is the product spectrum of PVC which focuses on durable products for the building industry. PVC products in domestic refuse are mainly packaging (65.6 %), consumer articles (23 %) and shoes (11.4 %).

TABLE 5.3
PVC products in domestic refuse (FRG 1985) according to Brahms et al. [4]

PVC products	Quantity in kt
Packaging	73.0
- blow mouldings (32.7 kt)	
- cups (18.9 kt)	
- moulded parts (15.1 kt)	
- films (6.2 kt)	
Consumer articles	25.6
·Shoes	12.7
Total	**111.3**

About 1.4 kg PVC residues accrue per head of population per year. Due to the small quantity and the difficulty to separate PVC from other plastics, the separate collection of PVC from domestic refuse is not an attractive proposition. The only alternative is to collect the plastics fraction in toto. PVC would amount to about 12 % of the plastics fraction.

TABLE 5.4
Plastics in domestic refuse

Annual Quantity	per inhabitant	Total
Polyethylene, polypropylene	6.8 kg	417,000 t
Polystyrene and others	3.4 kg	205,000 t
Polyvinyl chloride	1.4 kg	85,000 t
Total	**11.6 kg**	**707,000 t**

When collecting plastics from domestic refuse it probably makes sense to concentrate on larger, homogeneously composed products. Products of mixed composition like shoes or composite materials can hardly be utilized and films from domestic refuse would introduce a disproportionately high amount of impurities into the plastics fraction. The utilizable quantity would reduce to about 400 kt plastics with PVC products still accounting for around 12 %.

A large part of the plastics fraction of domestic refuse consists of hollow articles which are responsible for the extremely low bulk densities. Gallenkemper and Doedens [5] found the bulk density of plastics waste to be 28 to 46 kg/m^3 with 362 kg/m^3 being achieved only after pressing into bales. The low bulk density increases the transport cost. For comparison: the bulk density of waste paper in deinking quality is 157 - 280 kg/m^3, pressed into bales it is 602 kg/m^3.

The low bulk density, the high level of contamination of the plastics and the lack of type purity are problems affiliated with collecting plastics waste from households [5]. Using storage containers only 1.25 kg of Plastics per inhabitant and year have been collected in Hamburg. Better collection results were obtained with the bag system, which gave yields of 6 kg per inhabitant and year. Gallenkemper and Doedens estimated the costs of collection in storage containers at 400 - 700 DM/t plastics. In pilot projects employing various systems the costs fluctuated from 300 to 2000 DM. Average costs for container collection in Erlangen [6] were 1,270 DM/t plastics. Table 5.5 shows how these costs were derived. Erlangen also experimented with siting of "Green bins". These gave rise to costs of 1,664 DM/t plastics.

TABLE 5.5

Costs of container collection of plastics from domestic refuse in Erlangen (1985/1986) [6]

Nature of cost	DM/t
Transport to intermediate storage	68
Transport to recycling plant	244
Container rentals	53
Pressing (into bales etc)	61
Weighing cost	4
Discharge, collection	653
Intermediate storage ports	63
Cost of recycling	124
Total	**1,270**

5.4 Contribution of PVC to the harmful substances content of municipal waste

In the Federal Republic there is an annual accumulation of around 29 million t municipal waste. This waste consists of domestic refuse, commercial waste similar to domestic waste, bulky waste, road sweepings and refuse from markets. Some 30 % of this municipal waste is incinerated, while most of the remainder is landfilled. In 1987, municipal waste contained about 170 kt PVC, corresponding to 0.57 % by weight. The PVC content of the municipal waste originated to the extent of 50 % from domestic refuse (85 kt), 41 % from commercial waste (70 kt) and 9 % from other provenances, mainly bulky waste (15 kt).

Table 5.6 compares the harmful substances content of the PVC fraction in municipal wastes with the total of these substances in municipal waste. For this purpose the data published by Reimann [7] were taken for the composition of the municipal waste. The heavy metal content of the PVC fraction of domestic refuse was investigated by Brahms, Eder and Greiner [4]. The heavy metal content of PVC in commercial waste and in bulky waste had to be estimated. It is assumed that these waste streams contain the aliquot part (85 kt PVC waste correspond to 8 % of the polyvinyl chloride processed in the country) of the heavy metal used in the formulation of PVC compounds (see Chapter 4). Table 5.6 shows the contribution made by PVC to loading waste with harmful substances. The figures approximately reflect the situation in 1985. If new data were available, one would reckon with a decidedly lower cadmium loading from PVC products.

TABLE 5.6

Contribution made by PVC products to the harmful substances content of municipal waste
(tonnes, FRG 1985)

Element	PVC contribution			Total content of waste	PVC fraction of this
	Domestic waste	Other waste	Total waste		
Chlorine	48,169.5	48,169.5	96,339.0	217,500	44.3 %
Cadmium	11.0	22.2	33.2	304	10.9 %
Lead	46.6	737.0	783.6	13,050	6.0 %
Zinc	19.5	10.0	29.5	43,500	0.1 %
Chromium	4.7	15.0	19.7	7,250	0.1 %
Copper	5.9	5.0	10.9	14,500	0.1 %
Nickel	0.7	1.0	1.7	2,320	0.1 %

Municipal waste contained 220 kt chlorine. The contribution made to the chlorine content by PVC wastes was 45 %. The contribution made by PVC to the cadmium content of domestic waste was 11 t, which originated from consumer objects [4]. Commercial waste and other sources are likely to have contained somewhat more than 20 t cadmium. Thus for 1985, 10.9 % of the cadmium content of municipal waste has to be attributed to PVC. As consumer goods are no longer stabilized with cadmium, the cadmium content of PVC products in domestic waste should be smaller today than in 1985. Lead stabilization also is of secondary importance for products having an impact on domestic refuse. In 1985, the PVC fraction in domestic waste contained 46.6 t lead. PVC packaging barely makes a contribution to the lead load (packaging 4.5 t, consumer objects 42.1 t). As made evident by the chromium content (4.7 t) of the PVC fraction in domestic waste, a part of the lead loading can be traced back to the use of lead chromate containing pigments (chrome yellow, molybdate red). The lead content of PVC waste from other sources would be higher. Taking a proportionate amount of the lead consumption for stabilization, these waste streams are likely to contain 677 t lead. If one estimates the entry of lead chromate pigments at 60 t, then the lead loading of municipal wastes caused by PVC products is of the order of 783 t, or 6 % of the total lead loading. Zinc and tin stabilization (32.6 t Sn in PVC from domestic waste) is frequently used in PVC packaging. The nickel and copper content of the PVC fraction is attributed to the use of pigments (nickel/titanium or phthalocyanine pigments). However, the part played by PVC in loading municipal wastes with zinc, tin, copper and nickel can be neglected.

References

1. Produktion im produzierenden Gewerbe des In- und Auslands. Fachserie 4, Reihe 3.1, Published by: Statistisches Bundesamt, Wiesbaden, 1967 - 1989.

2. Außenhandel nach Waren und Ländern (Spezialhandel). Fachserie 7, Reihe 2, Published by: Statistisches Bundesamt, Wiesbaden, 1975 - 1989.

3. Abfallbeseitigung im Produzierenden Gewerbe und in Krankenhäusern 1984 Fachserie 19, Reihe 1.2, Published by: Statistisches Bundesamt, Wiesbaden, 1987.

4. Brahms, E., Eder, G., Greiner, B., Papier - Kunststoff - Verpackungen Mengen und Schadstoffbetrachtung. BMFT-Forschungsbericht 1430368, 1988.

5. Gallenkemper, B., Doedens, H., Getrennte Sammlungen von Kunststoffen des Hausmülls. E.Schmidt Verlag, Berlin, 1988.

6. Kempe, K.H., Sammlung von "Alt"-Kunststoffen aus der Sicht einer Kommune. In Stoffliche Verwertung von Abfall- und Reststoffen, VDI Bildungswerk, 1989.

7. Reimann, D.O., Schwermetalle und anorganische Schadstoffe im Hausmüll mit ihrer Verteilung auf die feste und gasförmige Phase. VGB Krafwerkstechnik 1988, 68, 837 - 841.

6. Recycling of PVC waste

6.1 Introduction

The same recycling methods used for other thermoplastics, apply to the reclamation of PVC waste [1]. Clean, single material can be ground to produce recyclates which are able to replace prime PVC in many applications. The high density of PVC also makes a separation from plastics mixtures possible [2]. Plastics mixtures containing PVC can be processed to thick-walled products without separation [3]. If the plastics are highly degraded or heavily soiled, they can no longer be processed as thermoplastics. Heating in the absence of air (pyrolysis [4]) or reaction with hydrogen (hydrocracking [5]) destroys the macromolecules. This processes yield products similar to crude oil which can be used as chemical raw materials or energy source.

The recycling products are extremely diverse - the spectrum ranges from high value recyclates to energy sources. It is therefore difficult to compare the effectiveness of different recycling processes. However, it is possible to compare the energy equivalent of these products. The energy equivalent of PVC is 53 GJ/t [6]. This takes into account the energy content of naphtha, the raw material, the necessary process heat and the electric energy for the chlor-alkali electrolysis and for the operation of the plants. For comparison, the energy equivalent of 1 kWh of electric energy is 10.2 MJ (power plants only utilize 38 % of the calorific value of fossil fuels) [6]. The energy equivalent of crude oil is 43.3 GJ/t.

Single material, clean PVC residues are ground to produce a recyclate which can replace prime PVC in many uses. To grind 1 t PVC scrap requires 25 to 250 kWh electric power [7], depending on the particle size required. This corresponds to an energy equivalent of 0.25 to 2.5 GJ. So in the most adverse case less than 5 % of the energy expenditure needed for the manufacture of PVC is needed for recycling. If, in addition to grinding, the residues has to be washed, separated from other plastics in a hydrocyclone, dried and granulated, the power requirement is 700 kWh per tonne plastics [8] or, in round figures, 7 GJ expressed as energy equivalent. Thus slightly more than 10 % of the energy equivalent of PVC has to be employed to obtain a product which can partially replace virgin PVC. Recycling products from the thermoplastic processing of mixed plastics have too low a mechanical load bearing capacity to act as replacement for prime plastics. They compete much more with such materials as wood or concrete [9]. If concrete is replaced by such recyclates there is a saving of about 3 GJ energy which would otherwise be required for the manufacture of the clinker. Thus only 5 % of the energy equivalence value of PVC is utilized.

Hydrocracking of plastics waste gives rise to a hydrocarbon mixture which can be used as an energy source or as chemical raw material. One part of the hydrocarbon mixture has to be used for the provision of hydrogen. If 1 t PVC is hydrocracked, 300 kg of a hydrocarbon mixture with an energy equivalent of around 12 GJ remain. In the pyrolysis of PVC only small amounts of volatile hydrocarbons are formed apart from hydrogen chloride and carbon black. This procedure barely utilizes the resource offered by PVC waste. In waste incineration a part of the energy content of PVC is recovered in the form of electric power. One tonne PVC gives rise to 900 kWh power whose energy equivalent (9 GJ) is 15 % of the energy equivalent of PVC.

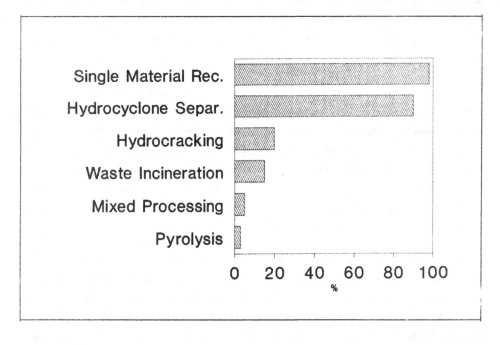

Figure 6.1:   Exploitation of the useful material/recoverable energy content of PVC on reprocessing

Figure 6.1 summarizes these values. The energy equivalent of PVC was taken as 100 %. The figure impressively shows the superiority of methods which produce a single material recyclate. It stresses the importance of arranging the collection of residues in such a way that the largest possible amounts of high quality and single material scraps are obtained.

Conservation of finite resources is but one goal of recycling. Landfill sites are becoming increasingly scarce too. In the thermoplastic processing of PVC residues, secondary waste hardly occurs. Only impurities have to be separated and disposed of. In all destructive processes (hydrocracking, pyrolysis, incineration) hydrogen chloride is formed as the main product. Hydrogen chloride is retained in a scrubber and usually neutralized. If this is done with sodium hydroxide, sodium chloride is obtained which can be reused in chlor-alkali electrolysis. Calcium hydroxide is much cheaper than sodium hydroxide. The neutralization product calcium chloride, however, cannot be sold and has to be run into the effluent or dumped. It is also possible to distil the scrubber liquid and obtain marketable hydrochloric acid.

Further residues of destructive utilisation are created by inorganic plastics additives. While only a few percent of the original weight are involved, careful disposal is necessary because of the heavy metals content. It is possible to reclaim cadmium and lead from the filter dusts. The pyrolysis of plastics additionally produces large quantities of soot which have to be disposed of in a separate process.

6.2 Single material PVC residues

Single material plastics waste occurs at the manufacturers of semi-finished stock as sprues, as faulty batches and on starting up and shutting down the processing machines. Cutting and stamping also yields high quality residues. Single material residues from domestic waste can only be obtained by means of preliminary separation. The number of process steps necessary for recycling depends on the quality of the waste. As a general rule these steps involve

- Removal of foreign substances
- sorting into colours
- coarse grinding
- fine grinding
- air classification

Fine grinding can be omitted, if a somewhat lower surface quality can be tolerated. If the converter processes PVC in powder form, the ground stock can be added directly to prime PVC. If the recyclate is to be sold, it is usually granulated. A recyclate content of 10 to 15 % does not cause any deterioration in quality even in highly stressed parts such as window profiles. Coextruded window profiles can accommodate up to 60 % recyclate. Far higher proportions of recyclate can be employed in PVC products used for interior or concealed applications. The PVC compounds usually contain sufficient reserves of stabilizer. The addition of stabilizer thus is rarely necessary.

The costs for recycling profile scrap are 200 to 300 DM/t. Investment costs are low. A mill which, depending on the material, is capable of processing 0.5 to 1 t PVC per hour, would cost around 65,000 DM. This can be done in small units. The energy requirement for grinding is in the region of 25 to 50 kWh per tonne PVC [7]. This increases to 100 to 250 kWh/t, if particle sizes below 0.8 mm are needed [10]. 250 kWh correspond to an energy equivalent of 2.5 GJ or 5 % of the energy required for the manufacture of PVC (53 GJ/t).

6.3 Solvent extraction

PVC dissolves in organic solvents such as tetrahydrofuran or cyclohexanone and may be precipitated from these solutions by addition of polar solvents such as methanol or ethanol [11,12]. This method suggested itself for the reclamation of composite materials and PVC coatings where a part of the structure is formed by a backing material which would preclude further thermoplastic processing.

Tetrahydrofuran and cyclohexanone have proved themselves the most suitable solvents. PVC is precipitated by addition of methanol or ethanol. After drying, the precipitated PVC is hard and granular. Projecting forward from gram scale laboratory experiments, about 10 m^3 THF are required for extraction of one tonne PVC and 12 m^3 ethanol for its precipitation. The plasticizer is partly in the precipitated PVC and partly in the solvent mixture. The solvents have to be recovered by distillation. Kampouris et al. [12] indicates a 90 to 95 % yield on rectification which means a solvent and precipitator loss of 1 to 2 m^3 per tonne recovered PVC. The procedure hence is costly and ecologically fairly pointless.

6.4 Separation of mixed plastics scrap

Electrostatic processes, flotation, optical methods and density separation were suggested for the separation of mixed plastics scrap. Electrostatic processes are impracticable of the small difference in the surface conductivity of plastics and the considerable influence impurities have on the separating effect [2]. The separating principle of flotation depends on the attachment of air bubbles to solid particles suspended in water. By addition of suitable wetting agents PVC-U can be separated from polystyrene, but the behaviour of plasticized PVC is non-specific [2]. Optical methods are employed for the separation of PVC and polyethylene terephthalate. This process is mainly of interest in countries where PET as well as PVC bottles are common.

From a technical aspect the most important separating system exploits the density difference between the common plastics [2,8,13]. This difference is shown in Table 6.1. In the case of polyvinyl chloride, the density depends mainly on the plasticizer content. Pure PVC has a density of 1.39 g/cm^3.

Separation may be effected by a sink-float process: In the first stage the polyolefins are separated from polystyrene and PVC by floating in water. In stage two, polystyrene is separated from PVC by floating in an aqueous solution of $CaCl_2$ (density 1.07 g/cm^3). The amount which can be put through is limited by the need to avoid turbulence in the separation region.

TABLE 6.1
Density of the common plastics [14]

Plastic	Abbreviation	Density (g/cm^3)
Low density polyethylene	PE-LD	0.91
High density polyethylene	PE-HD	0.95
Polystyrene	PS	1.03
Polyvinyl chloride	PVC	1.19 - 1.37

Far better throughput performances can be achieved in a hydrocyclone. Due to the centrifugal acceleration and the geometry of the hydrocyclone the throughput performance is 100 times greater than in a static sink-float separator of comparable size [2]. The Technical University Clausthal carried out trials using a hydrocyclone with a capacity of 1 t/h [2]. The plastics used were separated from domestic waste. The amount of impurities was 1 to 5 % (paper, metal, sand). The average composition was 80 to 85 % PE, 2 to 10 % PS and 8 to 15 % PVC. Impurities were separated using a vibrating screen. A two stage hydrocyclone was used for the actual separation.

TABLE 6.2
Separation of plastics waste in a laboratory scale hydrocyclone [2]

Starting material: plastics from domestic waste, particle size <15 mm	
Polyethylene	800 - 850 kg
Polystyrene	20 - 100 kg
PVC	80 - 150 kg
Impurities	10 - 50 kg
Product: 3 plastics fractions	Yield
Polyethylene, (content 97 %)	100 %
Polystyrene	50 %
PVC (content 99.9 %)	98 %
Impurities	100 %

In the first stage a polyethylene fraction was obtained which still contained about 3 % polystyrene. Practically 100 % of the polyethylene was removed. The underflow consisted of polystyrene and PVC and was separated in the second stage in a flat bottom cyclone. The PVC, which was thus received, had a purity of 99.96 %. Table 6.2 shows the result of separation achieved in this hydrocyclone.

Commercial plants currently only separate in a light fraction (polyolefins) and a heavy fraction (PVC, Polystyrene) [8,13]. Only the light fraction can be sold. The following operations are carried out:

- breaking up bales of plastics scrap into small pieces
- removal of coarse mineral impurities
- wet grinding
- separation in a hydrocyclone
- dewatering
- thermal drying
- intermediate storing and homogenizing
- granulating
- drying and storing.

The process cost is around DM 700 per tonne. The investment costs for a plant with a throughput of 1 t polyethylene per hour are around 5.3 million DM [8]. The energy requirement is around 700 kWh per tonne regranulated material, the fresh water requirement is 10 m^3 per tonne [8]. A power requirement of 700 kWh corresponds to an energy equivalent of slightly greater than 7 GJ. This is somewhat more than 10 % of the energy required for the manufacture of the plastic.

6.5 Recycling of mixed plastics scrap without separation

Mixed plastics scrap can be processed to thick walled articles even without prior separation. These processes also tolerate a certain amount of materials other than plastics. Inherent plastics properties severely restrict the usability of goods from mixed plastics.

As shown in Figure 6.2, common mass produced plastics differ widely in their processing temperatures. There is no temperature range which is optimum for all types of plastics. PVC in particular is processed at very low temperatures within a narrow range.

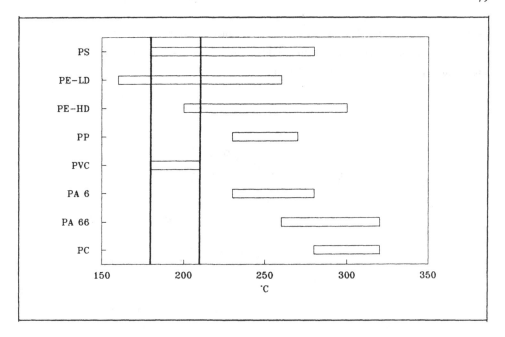

Figure 6.2:   Processing temperatures of large volume plastics according to Schönborn [16]

Because of the limited compatibility of the plastics, the polymer mixture does not form a homogeneous phase on setting. The mechanical properties of the recyclate are decidedly worse than the values shown by the single material plastics. It is therefore possible to produce only thick walled articles which are atypical for plastics.

Although more or less any mixture of plastics can be processed, operators of most plants switched to processing essentially single material plastics scrap [16]. This was necessary in the interest of a mainly trouble-free operation and homogeneous product quality. Additionally, highly contaminated plastics led to severe machine wear and markedly lower product quality. In principle, there are two methods for processing mixed plastics scrap [17]:

-   Thick bars and profiles can be manufactured on robust short screw extruders.
-   Thick walled moulded parts are manufactured by plasticizing the residues on a roller extruder followed by compression moulding.

The price of plants can differ markedly. Plants on which plastics scrap is broken down into small particles, plasticized on a roller extruder and subsequently moulded by means of hydraulic presses, are very expensive. Investment costs for plants with a throughput between 0.5 and 1 t recyclate per hour are of the order of 9 million DM [15]. The energy requirement is around 400 to 800 kWh/t product. Due to the high investment cost, manufacturing costs can be up to 1000 DM/t. Processing costs in the short screw extruder are decidedly lower, around 400 to 500 DM.

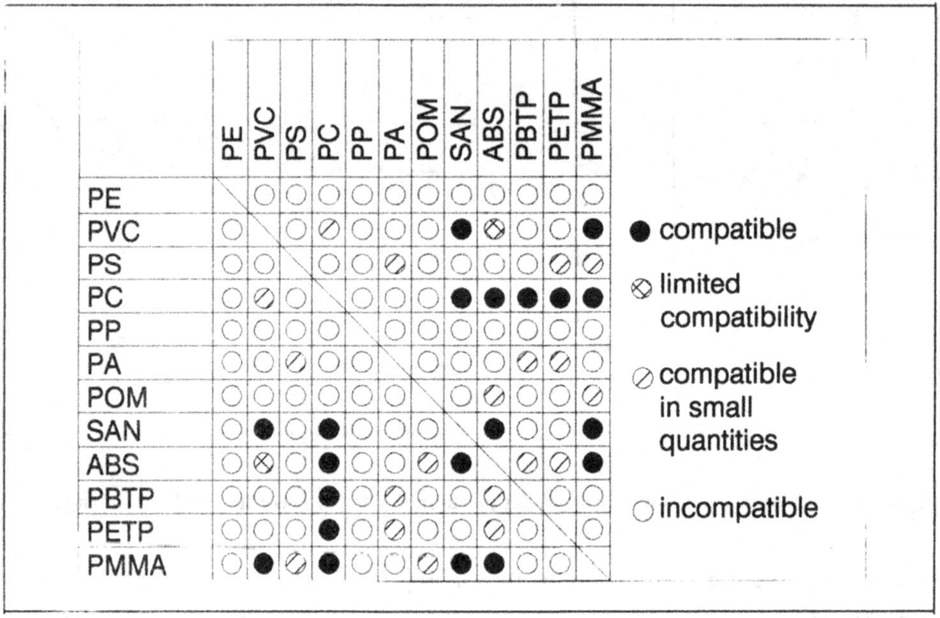

	PE	PVC	PS	PC	PP	PA	POM	SAN	ABS	PBTP	PETP	PMMA
PE		○	○	○	○	○	○	○	○	○	○	○
PVC	○		○	⊘	○	○	○	●	⊗	○	○	●
PS	○	○		○	○	⊘	○	○	○	⊘	⊘	⊘
PC	○	⊘	○		○	○	○	●	●	●	●	●
PP	○	○	○	○		○	○	○	○	○	○	○
PA	○	○	⊘	○	○		○	○	○	⊘	⊘	○
POM	○	○	○	○	○	○		○	⊘	○	○	⊘
SAN	○	●	○	●	○	○	○		●	○	○	●
ABS	○	⊗	○	●	○	○	⊘	●		⊘	⊘	●
PBTP	○	○	⊘	●	○	⊘	○	○	⊘		○	○
PETP	○	○	⊘	●	○	⊘	○	○	⊘	○		○
PMMA	○	●	⊘	●	○	○	⊘	●	●	○	○	

Legend:
- ● compatible
- ⊗ limited compatibility
- ⊘ compatible in small quantities
- ○ incompatible

Figure 6.3:   Blendability of plastics [18]

The main problem in mixed plastics processing is the quality of the products, as only thick walled products can be produced. There is only a limited market for such products, estimated to be some at 10,000 t/a [18]. To expand sales, the recyclates have to capture markets which are not the domain of plastics [16]. They have to compete with cheap materials such as wood and concrete [9]. One tonne recyclate replaces around 2 t concrete for whose manufacture 10 GJ energy have to be expended. But for processing the recyclate, some 7 GJ energy is required, so that the use of recyclate in place of concrete saves about 3 GJ energy.

TABLE 6.3
Properties of recycled mixed plastics in comparison with prime plastics [16]

Property	Replast I	Replast II	PE-LD	PE-HD	PVC	PS-HI
Modulus of Elasticity (Rigidity, N/mm^2)	850-1250	600	500	1300	3000	3000
Yield strength (Tensile strength,N/mm^2)	16-19	15	10-15	20-23	40-60	20-40
Elongation at break (%)	3-6	13	600	800	10-50	40-80
Notched impact strength (kJ/m^2)	2-4	8	n.b.	6-n.b.	3-5	4-6

Replast I: Plastics from domestic waste   Replast II: Recyclate from PE-LD films
n.b. = no break

6.6 Pyrolysis of PVC containing plastics scrap

The concept of pyrolysis is thermal decomposition under exclusion of air, with the production of pyrolysis oils and gases. The products can be used as a chemical raw material or as a fuel [4]. Products of pyrolysis are a heating gas, which has to be partly burned to maintain the pyrolysis, and oily hydrocarbon fractions similar in value to crude oil. The yield of heating gas and pyrolysis oils depends markedly on the composition of the starting material. Additional substances formed are process water, which has to be removed, a tar fraction, which also contains carcinogenic constituents [19], soot and inorganic residues. Because of the current low crude oil prices, the collection and transport costs alone exceed the return on the products by several times. The process particularly promoted in Germany was fluidized bed pyrolysis. The University of Hamburg has an experimental plant operating on a kg/hour scale. A demonstration plant in Ebenhausen has a throughput of 0.3 to 0.5 t/h.

Under the conditions of pyrolysis PVC forms hydrogen chloride. This has to be bound by the addition of dolomite. In the current plant this leads to breakdowns due to clogging of the heating gas line [19]. Alternatively, a separate dehydrochlorination at low temperature (approx. 300 °C) may be operated as an additional, upstream process step [19]. Dehydrochlorinated PVC has a carbon to hydrogen ratio of 1 to 1. The yield of volatile hydrocarbons is correspondingly low. A greater proportion of soot is formed here than in the pyrolysis of other plastics. This soot has to be disposed of in an incinerator. In the pyrolysis of plastics therefore, PVC contributes little to the formation of utilizable products.

6.7 Hydrocracking of PVC containing plastics scrap

In this process plastics waste is reacted with hydrogen under high pressure. As in pyrolysis utilizable products such as gases and oils are formed, but in decidedly higher yields [5]. The plastics waste is comminuted. Contaminants such as metals or glass are separated. The plastics particles are made into a mash with oil or other suitable substances and subsequently fed into the hydrocracking reactor. Conversion occurs at temperatures up to 500 °C and pressures up to 400 bar. Under these reaction conditions long chain molecules are broken and heteroatoms such as chlorine, sulphur, nitrogen and oxygen are split off. The hydrogen saturates the products. Afterwards all conversion products are fed to scrubbers where inorganic compounds are separated as neutral salts. The liquid reaction products are distiled to yield chemical base materials, petrol, heating oil and other products. The volatile hydrocarbons are used as heating gas [5]. One tonne plastics scrap from municipal waste (cleaning bottles, yoghurt pots, etc.) will generate 650 kg oil products and 170 kg heating gas [20]. The residue (180 kg) contains pigments, fillers, additives and mineral constituents.

Under the conditions of hydrocracking the PVC chains are broken and saturated. Chlorine is split off as hydrogen chloride. As an admixture in other plastics wastes, PVC presents no problem for hydrocracking [20]. According to stoicheiometry, PVC mainly gives rise to hydrogen chloride and fewer hydrocarbons. The empirical equation approximates to:

$$(C_2H_3Cl)_n + 1.2 \, n \, H_2 \; = \; 2n \, CH_{2.2} + n \, HCl$$

Calculation of yields was based on a hydrocracking product having the approximate composition $CH_{2.2}$. Hydrocracking of 1 t PVC gives rise to 584 kg HCl and 454 kg hydrocarbons. To provide the hydrogen (38.4 kg) about 150 kg of hydrocarbons have to be steam reformed. The hydrocracking reaction is exothermic. Power requirement for operating the plant was not taken into account for the purpose of this consideration. There remain 300 kg of a hydrocarbon fraction with a lower heating value of 40 MJ/kg. This corresponds to an energy equivalence value of 12 GJ, which is about 20 % of the energy equivalence value of PVC.

Investment costs for a plant hydrocracking 100,000 t plastics waste per year are of the order of 100 million DM [21]. At current low oil prices capital cost alone would surpass the return which could be expected from selling the hydrocarbons. On top of this are the cost of collection, transport and the operating costs. The yield of useful materials is greater for the hydrocracking of polyolefins than of PVC, although hydrocracking is uneconomic even in the most favourable case.

6.8 Waste incineration

Incineration of waste utilizes its combustion energy. As an additional benefit the weight and volume of wastes is strongly reduced. In Germany at present, about one third of municipal waste is incinerated. This waste contained 57,000 t PVC.

In chapter 5 it was attempted to estimate the contribution made by PVC products to the heavy metal load of municipal waste. The result was summarized in table 5.6. One third of this quantity goes into municipal solid waste incinerators. To demonstrate where the pollutants remain, these values are correlated with data on the mass stream of pollutants in waste incineration plants. Reimann compiled the values for pollutant flow in a plant with dust separator, single stage flue-gas scrubber and slag washer [22].

TABLE 6.4

Pollutant remains on combustion in waste incineration plants. Figures in % according to Reimann [22]

Element	Emission (stack)	Slag washed	Filter dust	Flue-gas wash water	Slag wash water
Chlorine	3.0	8.0	15.0	67.0	7.0
Cadmium	2.0	10.0	85.0	3.0	0.5
Lead	0.7	67.0	32.0	0.7	0.1
Zinc	0.2	73.0	26.0	0.8	<0.1
Chromium	<0.1	94.0	6.0	0.2	<0.1
Copper	<0.1	93.0	7.0	0.1	0.1
Nickel	<0.1	93.0	5.0	2.0	<0.1

If these percentage figures are transferred to the pollutants introduced into waste incineration plants via PVC products, one obtains a picture of where the pollutants remain: With PVC waste 32 kt chlorine were introduced into waste incineration plants. Of this quantity about 1000 t were emitted and 2,500 t were fixed in the slag. The remainder was located in the flue-gas wash water and in the slag wash water. Incineration of PVC residues caused 0.2 t Cd emissions. This corresponded to less than 1 % of the total cadmium emissions into the air (sum of all cadmium emissions into the air in the FRG in 1986: 35 t [23]). Waste incineration of PVC caused an emission of 1.6 t lead. The largest part of lead remained in the slag and in the filter dust. In the Federal Republic of Germany a total of some 4000 t lead was emitted into the air in 1986. 3200 t lead emissions resulted from the use of leaded petrol. Most of the remainder came from metal production and from power stations [23]. Thus the incineration of PVC waste contributed decidedly less than 0.1 % to the lead pollution of the air in the FRG. The emissions of zinc, chromium, nickel and copper, which are attributable to PVC, may be neglected, partly because PVC products hardly contribute to the metal content of waste and partly because of the low volatility of these elements.

TABLE 6.5

Remains of pollutants from PVC products on combustion in waste incineration plants (in tonnes)

Element	Load due to PVC	Emission (stack)	Slag washed	Filter dust	Flue-gas wash water	Slag wash water
Chlorine	32,113.0	963.4	2,569.0	4,817.0	21,515.7	2,247.9
Cadmium	11.1	0.2	1.1	9.4	0.3	0.1
Lead	241.3	1.6	161.7	77.2	1.8	0.2
Zinc	9.8	<0.1	7.2	2.6	0.1	<0.1
Chromium	3.2	<0.1	3.0	0.2	<0.1	<0.1
Copper	3.6	<0.1	3.4	0.3	<0.1	<0.1
Nickel	0.6	<0.1	0.5	<0.1	<0.1	<0.1

The large contribution made by PVC residues to the chlorine load in municipal waste (45 %) suggests that PVC plays a part in the formation of chlorinated dibenzodioxins and furans. By varying the waste composition an attempt was made to prove or refute the effect of PVC on the formation of these compounds. In this work no proof could be found of any significant influence by the PVC contents in waste on the production of PCDD and PCDF [24]. In 1984 tests have been carried out on a French incineration plant using waste to which PVC and chlorophenols had been added [25]. The PVC content had been increased by 300 %, nevertheless, the dioxin content of the ash virtually remained unaltered by these additions. Similar tests and determinations were also carried out at the municipal waste incinerator in Berlin-Ruhleben [26]. The addition of PVC leads to an increase in the HCl content of the crude gas, but not to a significant increase in the PCDD/PCDF values. Even in the incineration of a low chlorine content fuel such as waste paper, the same quantities of dioxins and furans are formed as in the incineration of domestic refuse containing PVC [27]. Apart from the carbon content, the chloride content of flue dust is responsible for dioxin formation. This appears to be a function of the chemical composition (alkalinity) of the flue dust and is not, or only faintly, correlated with the HCl crude gas concentration [28].

After incineration, the chloride is present in aqueous solution in the flue-gas washer. If this solution must not be discharged into the effluent, the chloride has to be recovered as its calcium or sodium salt. On neutralizing the HCl content of 1 t PVC (pure polymer) it gives rise to 0.88 t calcium chloride or 0.94 t sodium chloride. Calcium chloride, which already arises in enormous quantities (approx. 1 million t p.a. in the FRG) in the production of sodium carbonate (sodium carbonate is predominantly used as feedstock in glass manufacture), cannot be placed on the market and has to be dumped. The landfill costs vary widely, but they can be expected to increase. This is the reason why at the waste incineration plants at Iserlohn and Stapelfeld the washing liquor is neutralized with caustic soda and evaporated [29]. The recovered sodium chloride is used in chlor-alkali electrolysis once more. The benefit of the process lies less in the value of the recovered sodium chloride, than in the landfill space which is saved. An interesting alternative is the recovery of hydrochloric acid from the flue-gas washer by distillation. This avoids the use of costly sodium hydroxide.

The calorific value of PVC is 18 GJ/t, decidedly higher than for the average for the waste (8.5 GJ) [6]. The calorific value of plasticized PVC is higher because of the plasticizer content. Friege et al. [30] point out that the throughput of a waste incineration plant is limited by the calorific value of waste and that too high a plastics content would reduce the capacity of existing MSW incinerators. Nevertheless, the contribution made by PVC to the average calorific value is so low, that PVC has hardly any significant effect on the throughput of the waste incineration plants: The calorific value of PVC amounts to 1.5 to 2 % of the calorific value of municipal solid waste. Reimann found that due to the PVC content of municipal waste there is an annual saving in the use of heating oil of about 20,000 t [31].

In Germany the main product of waste incineration plants is electric power, sometimes it is possible to also sell district heating. The energy efficiency in waste incineration plants is not as good as in power stations and amounts to 18 %. This means that 51 GWh power is produced from 57,000 t PVC (neglecting the energy content of plasticizers). One tonne PVC generates 900 kWh power which has an energy equivalence value of 9.2 MJ. Thus, by the incineration of waste it is possible to recover 15 % of the energy equivalent.

References

1. Gaensslen, H., Sordo, M., Tötsch, W., Produktion, Verarbeitung und Recycling von PVC. Fraunhofer-Institut für Systemtechnik und Innovationsforschung, April 1989.

2. Bahr, A., Vogt, V., Djawadi, H., Sortierung von Kunststoffabfällen. Report of the Research Program "Wiederverwertung von Kunststoffen", Vol. 4, TU-Clausthal, 1980.

3. Schönborn, H., Verwertbarkeit von Kunststoffen. In Konzepte zur Gewinnung von Wertstoffen aus Hausmüll. Mülltechnisches Seminar, TU München, 1986.

4. Kaminsky, W., Sinn, H., Pyrolyse von Kunststoffabfällen. Report of the Research Program "Wiederverwertung von Kunststoffen", Vol. 5, Universität Hamburg, 1980.

5. Löffler, W., UK-Wesseling-Verfahren zur Umwandlung von Kunststoff zu Öl. In Stoffliche Verwertung von Abfall-und Reststoffen. VDI Bildungswerk, 1989, pp. 121 - 143.

6. Kindler, H., Nikles, A., Energieaufwand zur Herstellung von Werkstoffen - Berechnungsgrundsätze und Energieäquivalenzwerte von Kunststoffen. Kunststoffe 1980, **70** 802 - 807.

7. Leschonski, K., Goritzke, W., Röthele, Aufbereitung von Kunststoffabfällen zum Zwecke der Wiederverwertung (Zerkleinern, Klassieren). Research Program "Wiederverwertung von Kunststoffabfällen", April 1981.

8. Stolzenberg, A., Bewährte Technik im neuen Prozeßdesign. In Recycling von Abfällen 1, ed. Thomé-Kozmiensky, EF Verlag, Berlin, 1989.

9. Herrler, J., Ist ein umweltverträgliches Recycling von Kunststoffabfällen aus dem Hausmüll möglich? Müll und Abfall, **1986**, 238-242.

10. Herbold, K., Feinmahlen von Kunststoffabfällen löst Qualitätsprobleme. Kunststoffe, 1987, **77**, 1141-1142.

11. Schwenkedel, S., Koschwitz, D., Wiedergewinnung von PVC aus PVC-beschichteten Abfällen. Kunststoffe 1980, **70**.

12. Kampouris, E.M., Diakoulaki, D.C., Papaspyrides, C.D., Solvent Recycling of Rigid Polyvinyl Chloride Bottles. Journal of Vinyl Technology, 1986, **8**, 79 - 82.

13. Konstantius, R., Kunststoffaufbereitung nach Andritztechnologie. In Recycling von Abfällen 1, ed. Thomé-Kozmiensky, EF Verlag, Berlin, 1989.

14. Saechtling, H., Kunststofftaschenbuch, Hanserverlag, München Wien, 1986.

15. Härdtle, G., Marek, K., Bilitewski, B., Kijewski, K., Recycling von Kunststoffabfällen. Beihefte zu Müll und Abfall, 1988, **27**.

16. Schönborn, H.-H., Alt- und Neukunststoffe - Übersicht. In Stoffliche Verwertung von Abfall- und Reststoffen, VDI Bildungswerk, 1989, pp. 57 - 75.

17. Mylenbusch, H., Zum Stand des Kunststoff-Recycling. In Recycling von Abfällen 1, ed. Thomé-Kozmiensky, EF Verlag, Berlin 1989.

18. Menges,G. et al., Recycling des Kunststoffanteils von PKW. Kunststoffe, 1988,**78**, 573 - 583.

19. Kaminsky, W., Kunststoffpyrolyse. In Stoffliche Verwertung von Abfall- und Reststoffen, VDI Bildungswerk 1989, pp. 145 - 158.

20. Written communication by Union Rheinische Braunkohlen Kraftstoff AG. 13.10.88.

21.   Löffler, W., UK-Wesseling-Verfahren zur Umwandlung von Kunststoffen zu Öl. In Stoffliche Verwertung von Abfall- und Reststoffen. VDI Bildungswerk 1989.

22.   Reimann, D.O., Schwermetalle und anorganische Schadstoffe im Hausmüll mit ihrer Verteilung auf die feste und gasförmige Phase. VGB Kraftwerkstechnik 1988, **68**, 837 - 841.

23.   Angerer, G., Böhm, E., Schön, M., Tötsch, W., Möglichkeiten und Ausmaß der Minderung luftgängiger Emissionen durch neue Umweltschutztechnologien. Fraunhofer-Institut für Systemtechnik und Innovationsforschung Bericht Nr. ISI-B-2-90, January, 1990.

24.   Jager, J., Wilken, M., Verminderung von Dioxinen und anderen Schadstoffen in Emissionen und Rückständen von Müllverbrennungsanlagen. In Müllverbrennung und Umwelt 3. ed. Thomé-Kozmiensky, EF Verlag Berlin, 1989, pp. 33 - 39.

25.   Gonnord, M.F., Measurements of Dioxins in Waste Incineration Plants. In Recycling International, ed. Thomè-Kozmiensky, EF-Verlag, Berlin, 1984.

26.   Sierig, G., Dioxinmessungen an einer Müllverbrennungsanlage. In Müllverbrennung und Umwelt. EF-Verlag, Berlin 1987, pp. 617.

27.   Martin, J., Zahlten, M., Betriebs- und Inputvariationsversuche an einer Müllverbrennungsanlage - Ergebnisse und Ausblick. Abfallwirtschaftsjournal 1989, **5**, 41 - 54

28.   Vogg, H., Hunsinger, H., Stieglitz, L., Beitrag zur Lösung des Dioxin-Problems bei der Abfallverbrennung. In Müllverbrennung und Umwelt 3. ed. Thomé-Kozmiensky, EF Verlag, Berlin, 1989, pp. 15 - 32.

29.   Kochsalz aus PVC. Kunststoffe-Plastics **1986**/11 29 - 30.

30.   Eberhardt, A., Friege, H., Schumacher, E., PVC in der Müllverbrennung. Müll und Abfall, 1986, **18**, pp. 377- 382.

31.   Reimann, O., PVC als Abfallprodukt. Müll und Abfall **1988** 256 - 267.

7. Current recycling activities

7.1 List of firms carrying out recycling

A list of firms processing PVC scrap is given on the following pages. The list was made available to us by the Arbeitsgemeinschaft PVC und Umwelt e.V. (AgPU). It originates from a survey carried out among the recycling companies [1]. The information was confirmed by the companies and is published with their agreement. Information is given on the nature of the scrap accepted (rigid, PVC-U, or plasticized PVC, PVC-P), its state (single material, mixed and/or soiled) and source (scrap from production or after service). If addresses are needed, they can be obtained from AgPU [1].

7.2 Postconsumer Recycling

Most of the commercial recyclers listed in Table 7.1 specialize in production waste. Postconsumer waste is much more difficult to recycle: It is inhomogeneous and usually contains lots of foreign material. The best strategy to obtain high quality recyclates from post consumer waste is as follows:

-    Identify a product group which is mainly manufactured from PVC (e.g. plastics windows, plastics pipes, floor coverings etc.)
-    Set up a collection scheme tailored to this product group
-    Build a specialized recycling plant
-    If possible, use the recyclates to manufacture the same product once more.

Recycling activities for postconsumer waste have been mostly initiated by PVC converters, who want to use the recycling scheme as a selling argument for their product. In most cases the PVC manufacturers participate in these recycling schemes, giving as well technical as financial assistance.

TABLE 7.1

PVC Recycling companies in West Germany

Type of PVC Origin of waste Condition (single mat. SM, mixed MX, soiled SD)		PVC-U Production			PVC-U Post consumer			PVC-P Production			PVC-P Post consumer		
		SM	MX	SD	SM	MX	SD	SM	MX	SD	SM	MX	SD
1000 Berlin 42	Firma S. Schulz	+	+						+				
1000 Berlin 47	BEAB GmbH Papier- & Kunststoff	+	+					+	+		+	+	+
1000 Berlin 51	BVB Kunststoffe	+	+	+	+			+			+	+	+
1000 Berlin 61	PAV Papieraufbereitungs & Verarb.	+	+	+	+	+		+					
2105 Seevetal 2	Recycling-Centrum Seevetal GmbH	+			+			+			+		
2170 Hemmor	RGH Recycling GmbH	+	+	+									
2955 Bünde	Kolthoff GmbH	+	+	+	+	+		+		+	+	+	+
2962 Großefehn 1	R. u. J. Beckmann	+						+		+			
3103 Bergen	KUVA Kunststoffverarbeitung	+											
3150 Peine	Plastobor Kunststofferzeug.	+	+	+	+		+	+	+	+	+	+	+
3258 Aerzen-Rehr	Sitte GmbH	+											
3380 Goslar	Nico-Metall GmbH								+	+	+	+	+
3446 Meinhard-Frieda	Friedola Gebr. Holzapfel												
3454 Bevern/Forst	Kajo Kunststoffe J. Stiegelhofer		+	+				+	+	+	+	+	+
3470 Höxter	Höku Kunststoffe	+											
4040 Neuss	Firma W. Lieth	+	+					+	+				
4060 Viersen 11	Hoffmann & Voss GmbH	+				+							
4100 Duisburg 14	Rockelsberg	+			+	+		+					
4150 Krefeld	Kunststoffe Kremer	+						+			+		
4330 Mühlheim a.d. Ruhr	Harry Jacob KG	+						+			+		
4418 Nordwalde	Plastikwerk Nordwalde GmbH	+							+				
4422 Ahaus	Excolo & Gelaplast .	+											
4424 Stadtlohn	Kunststoffverarb. GmbH Krumbeck	+											

Type of PVC	PVC-U Production			PVC-U Post consumer			PVC-P Production			PVC-P Post consumer		
Origin of waste												
Condition (single material SM, mixed MX, soiled SD)	SM	MX	SD	SM	MX	SD	SM	MX	SD	SM	MX	SD
4722 Ennigerloh — GEBA Kunststoffhandel & Rec.	+	+		+								
4763 Ense-Parsit — F. Pauli Metallwarenfabrik	+	+					+					
4788 Warstein — Monika Jäger Rohstoffhandel	+	+	+	+	+	+	+	+	+	+	+	+
4900 Herford — Schwerdt Kunststoffverwertung	+			+			+					
4930 Detmold-Piridsheide — Intra-Plast	+	+	+				+					
4952 Porta Westfalica — Porta Kunststoff-Recycling GmbH	+	+	+				+	+	+			
4973 Vlotho — IUH Porta Plastic GmbH	+	+	+				+	+				
5780 Bestwig — Stratmann Abfall-Recycling												
6050 Offenbach/Main — REKU GmbH Kunststoffrecycling	+	+		+	+		+	+		+	+	+
6124 Beerfelden/Odw. — Braun & Wettberg GmbH	+			+								
6361 Niddatal 3 — Klaus Pollak Kunststoff-Recycling	+			+	+		+					
6440 Bebra — Plus Plan Kunststoff & Verfahrenst.	+	+		+	+							
6520 Worms — WKR Altkunststoffprod. & Vertrieb.	+			+								
6530 Bingen 13 — Cogranu GmbH	+						+					
6800 Mannheim — Bormann Erben KG	+			+	+	+	+			+	+	+
6800 Mannheim 1 — GAS Gesellschaft für Abfallbes.	+	+		+	+		+	+				
6928 Helmstadt-Bargen 1 — Thermo-Plast Recycling	+											
7092 Rosenberg — Josef Rettenmeier + Söhne	+	+		+	+	+	+	+				
7253 Renningen 2 — Remax-Kunststoff	+	+	+	+	+							
7519 Eppingen — Cabka-Plast GmbH	+	+										
7944 Herbertringen — WOBA Kabel Verwertungs GmbH												
8000 München 50 — R. Bittl GmbH										+		
8070 Ingolstadt — Büchl GmbH & Co. KG							+					
8385 Pilsting — AVG Abfallverwertungsges.			+									
8437 Freystadt — Firma Manfred Leibold	+											
8663 Sparnek — R. Schöberl Chemierohstoffe	+	+		+			+			+	+	+
8785 Eussenheim — Sohler Plastik GmbH	+						+					
8890 Aichach — Spindler Kunststofftechnik	+						+			+		
8904 Friedberg 3 — Wagner Plastik GmbH	+			+			+			+		

PVC Floor coverings

Retailers and layers have always returned scrap from "as new" PVC floor coverings to the producers. A logistic system for returning scrap from used coverings is being established [2]. In April 1990 about 20 manufacturers and importers of PVC flooring and manufacturers of PVC set up the Arbeitsgemeinschaft PVC Bodenbelagsrecycling (AgPR) [3,4]. The goal of this association is to recycle the 100.000 t of PVC flooring, which is annually produced in Germany. Used PVC floorings are collected at the retailers, at do-it-yourself markets and by some municipal authorities. A recycling plant has been set up, where these scraps are processed to a fine powder, which is used for the manufacturing of new flooring.

The recycling plant is situated at Großefehn in the north of Germany. It has a capacity of 6,000 t/a. PVC floor coverings are separated manually from linoleum or carpeting. The PVC flooring is comminuted by cutting and grinding. The resulting powder is sold to the member companies of AgPR. New floorings are made from this recyclate by calendering. Because of the different colours of the materials used, this "raw" floor covering is usually dark beige or grey in colour. In the last stage of production a thin decorative film and a colourless wear-resistant film, both made from virgin PVC, are laminated on [2]. Thus the new PVC floor covering is made of up to 60 % of recycled material.

Window sections

With respect to logistics plastics windows are easier to handle than a good many other products. They are obtained either through renovation work or through the demolition of buildings. The association of window manufacturers has signed an obligation to accept used PVC windows and to recycle them. At present there are about 40 collecting points distributed all over Germany. Most of them are situated at important manufacturers of PVC windows [5].

PVC windows last 50 years or more. They are produced since the beginning of the sixties. Therefore relatively few PVC windows are returned. At present a fully automated dismantling plant would be uneconomic. Manual dismantling costs between 30 and 35 DM per window. Fittings and seals are removed, then the profiles are dismantled. The individual components - metal, glass and PVC - thus are separated. The milling of window sections is state of the art and has been practiced with production waste on a large scale. The resulting material is slightly discoloured. New window profiles can be made from this material by coextrusion: The core of the profile consists of recyclate. The surface is made from new material for purely aesthetic reasons.

Pipes

Some manufacturers of plastics pipes have obliged themselves to accept old pipes. The return, however, is very small: In 1990 only 150 t of pipes have been returned. Technically the recycling is very simple: The tubes are separated according to material, comminuted and washed. A further separation can be effected by means of a hydrocyclone. The material can be extruded without problems. An addition of auxiliaries is not necessary.

Bottles

In France, Operation Pelican was organized in 1989 to recover PVC from packaging in southern France. By paying 1200 FF/t this joint venture of GEOCOM and France Nature Environment hopes to collect 2000 t of PVC. A similar venture has been established recently among three resin producers and three mineral water suppliers in France. The price for the bottles has been set at 1350 FF/t. In the United States, a few vinyl industry pilot recycling projects are notable. In 1990, BF Goodrich began a program of collection of PVC bottles in Waukesha County, Wisconsin. Member companies of the Vinyl Institute started a collection program in the Philadelphia area [6].

The most difficult determination to make on a sorting line for plastics bottles is the difference between PVC and PET. There are a number of ways by which these two types of bottles can be distinguished. Extrusion blow moulded PVC has a horizontal scar from the parison pinchoff, while injection stretch blow moulded PET has a circular scar from the injection sprue. PVC typically has a slightly blue tint, whereas PET tends to be clear and colourless. Trivially, the recycling code designates PVC as "3". All of these methods, while accurate, do not square with the need for multiple identifications per second: Sorters work at a rate of a bottle per second or more. Typical samples of twice-hand-sorted bottles are about 95 % pure.

As a result, automated sorting technology is needed. PVC is the easiest material to identify instrumentally because it carries a marker: the chlorine atom. Many technologies have been developed to exploit this. Tecoplast Govoni, an Italian company, has patented a system which begins with reverse vending machines which accept bottles and subsequently flatten them between heated rollers. These slabs are transported to the sorting site and placed singularly on a cleated conveyer. Polymer identity is determined by transmittance of X-rays. ASOMA Instruments of Austin, Tx, utilizes X-ray fluorescence in a reflectance mode to identify singulated bottles passed near a detector. This device has been scaled up at the centre for Plastic Recycling Research at Rutgers, and the wTe recycling facility in Akron,

OH. National Recovery Technologies, Nashville, Tenn., has received grants from the Vinyl Institute and the United States Environmental Protection Agency (EPA) to develop a detection-ejection mechanism for PVC that does not require singulation. The first full-scale device has been built and will be installed at XL Disposal in Crestwood, Ill., in 1991. Recently other physical strategies for sortation of flakes of PVC from PET have surfaced. The REFACT hot-belt separator has been announced commercially. In this instrument, PVC melts and adheres to a heated surface. PET does not and can be removed easily from the surface [6].

Records

While compact discs are made from polycarbonate all conventional records are manufactured from PVC. The only problem for recycling arises from the labels which are made of paper. Therefore the inner part of the record is cut out. In Germany this work is performed by a workshop employing handicapped people [7]. The larger outer part of the record is comminuted and used in the manufacture of new records. The inner part is a mixture of paper and PVC. After milling this material can be used for less demanding applications.

Cables

In the past the reclamation of copper from cable scraps was an important source for PCDD/PCDF emissions: The scraps were incinerated in order to remove the plastics insulation material. In Germany this practice is discontinued. Nowadays the conductor and the plastics insulation is separated mechanically. Both materials are reused. The conductor can be reclaimed with a purity of up to 99,8 %. The plastics fraction consist mainly of PVC. It contains about 1 % metal. Nico-metal, a large converter of cable scraps, manufactures heavy duty flooring from the PVC fraction [8].

Other

Manufacturers of PVC roofing sheets have agreed to accept used PVC films [8]. Presently a site for a recycling plant for these roofing sheets is chosen. Also the wallpaper industry gives serious thought to the idea of recycling PVC wallpapers. These wallpapers consist of a paper backing and a PVC surface. It is possible to separate these materials by flotation. Collection schemes for medical articles are also probed. PVC packaging will have to be recycled according to the Verpackungsverordnung (ordinance on packaging waste) [9]. Here, however, PVC is just a small part of the general packaging recycling problem.

References

1.   Arbeitsgemeinschaft PVC und Umwelt. Adenauerallee 45, D-5300 Bonn 1

2.   Bonau, H.: PVC-Recycling of technical applications and economical aspects. Lecture held at: Recycle '90. Forum and Exposition, Davos (Switzerland), 29.05. - 31.05.1991

3.   Bathke, J., Eine Branche besinnt sich auf ihre Verantwortung. Boden Wand Decke 5, 1990

4.   Arbeitsgemeinschaft PVC Bodenbelagsrecycling. c/o Hüls Troisdorf AG, Kölnerstr. 176, D-5210 Troisdorf

5.   Verband der Fenster und Fassadenhersteller e.V. Bockenheimer Anlage 13. D-6000 Frankfurt a.M.

6.   Carrol, W.F., Vinyl recycling: an update. J.Vinyl.Technol. 1991, 13, 96 - 100

7.   Spindler, E. Wacker-Chemie GmbH, personal communication

8.   PVC-Recycling in der Praxis. Schriftenreihe PVC und Umwelt, Vol. 2. Arbeitsgemeinschaft PVC und Umwelt, May 1990

9.   Verordnung über die Vermeidung von Verpackungsabfällen (Verpackungsverordnung) of 14.11.1990

8. Summary

Polyvinyl chloride (PVC) is the oldest thermoplastic and, after polyethylene, still represents the most important of the general purpose plastics. In 1987, 1,320 kt PVC were manufactured in the Federal Republic. Taking account of the export trade surplus of PVC moulding compounds (243 kt), PVC semi-finished goods (160 kt) and PVC in finished products (approx. 100 kt), one is left with 800 kt PVC on the German home market (all calculations refer to the pure polymer).

The economic and technical success of this material is due to the attractive price and the outstanding material properties. These can be varied across a wide spectrum by adding different additives. PVC is mainly used in products for the building industry, examples being plastics windows, roller blinds, pipes, floor coverings, wall coverings and roofing sheets. In addition, PVC is used in electrical engineering (cable insulations, plugs/connectors, etc.), in the automotive industry (roof lining, underbody protection, tarpaulins for trucks), in the manufacture of gardening equipment (hosepipes), children's toys (balls and other inflatable toys, dolls), records, shoe-soles and in many other products. About 15 % of production are used for packaging. This includes blister packaging, bottles (e.g. for cooking oil, sun-tan products, mineral water, etc), packaging films and cups. In the medical field PVC products include blood bags, tubing and flow regulators for infusion and transfusion equipment.

Raw materials for the manufacture of PVC are sodium chloride and oil. Electrolysis of sodium chloride produces chlorine, caustic soda and hydrogen. It is the only technically important manufacturing process for caustic soda. Three processes are available for chlor-alkali electrolysis. They differ from each other in the way by which caustic soda and hydrogen, the cathode products, are separated from chlorine, the anode product. In the amalgam process, mercury forms the barrier. Sodium amalgam, which is formed at the cathode, is reacted with water in a separate decomposer to give hydrogen and caustic soda. In 1985, chlorine production by the amalgam process caused an emission of 5.5 t mercury; referred to 1 t PVC made with 'amalgam' chlorine, this amounts to mercury emissions of slightly less than 1.5 g. In the diaphragm process, a diaphragm separates the anode from the cathode compartment. Mercury emissions cannot occur, but the asbestos content of spent diaphragms makes it necessary for them to be disposed in special landfills. The membrane process is the most favourable, economically as well as ecologically. Chlorine and caustic soda are separated by an ion selective membrane. Neither mercury nor asbestos is involved in the process. The membrane consists of a perfluorinated plastic. The quality of the caustic soda is very high while the energy consumption is lower than in the other processes. So far no plant in the Federal Republic operates by the membrane process. Because of its many advantages, all new plants, will incorporate this process.

Chlorine reacts with ethylene to form 1,2-dichloroethane. This is cracked at high temperatures to give vinyl chloride and hydrogen chloride. Hydrogen chloride is either reacted with ethylene and oxygen to produce fresh 1,2-dichloroethane (integrated oxychlorination), or reacted with acetylene to form vinyl chloride directly (Hüls process). Both processes have yields of around 99 %. In the manufacture of vinyl chloride 20 to 30 t of the substance were emitted in 1987. When referred to 1 t PVC, the vinyl chloride emissions amount to approximately 20 g. Per tonne PVC, integrated oxychlorination gives rise to 34 kg by-products, while for the Hüls process the figure is 15 kg. These residues are either incinerated under controlled conditions, or are converted into chemical raw materials by reaction with chlorine.

PVC is produced by the polymerization of vinyl chloride. Three processes are available for this purpose: suspension, emulsion and mass polymerization. The resultant products differ in their uses and properties. In the suspension process the vinyl chloride and an initiator are finely dispersed in water by stirring. Product properties can be varied within a wide range. In the emulsion process the vinyl chloride is extremely finely dispersed in water with the aid of an emulsifier. Emulsion PVC is preferred for the manufacture of pastes. The mass process involves the polymerization of pure vinyl chloride and results in a very pure product.

Vinyl chloride is carcinogenic. It causes an extremely rare liver tumour, an angiosarcoma. Because this property was recognized too late, cancer of the liver occurred in workers exposed to high doses of vinyl chloride. World-wide, 140 cases of vinyl chloride induced liver tumour were identified. Once the risk was recognized, the production methods were drastically changed. Now plastics manufacturers produce PVC in totally sealed equipment in order to minimize exposure of workers to the substance. Exposure of workers has been reduced by a factor of 500 to 1000 since the mid sixties. The PVC sold must contain less than 10 g vinyl chloride per tonne. Frequently the vinyl chloride content is below 1 g/t. In order to achieve this low value, the polymer is stripped of monomer with large amounts of air. It is difficult to purify this quantity of air. In the Federal Republic about 300 t vinyl chloride was emitted in polymerization and monomer stripping in 1988. This corresponds to about 200 g vinyl chloride per tonne PVC. Vinyl chloride is rapidly decomposed in the atmosphere with a half life in the atmosphere of 2.4 days. In the processing of PVC, the vinyl chloride content of the polymer is reduced further to below 0.1 g/t. A maximum of 5 t vinyl chloride per year or 3.5 g/t PVC is liberated in processing PVC.

Apart from sodium chloride and the oil products ethylene or acetylene, the main requirement for the manufacture of PVC is energy. Energy is required to mine oil and rock salt, to transport the raw materials, for chlor-alkali electrolysis, as process heat and for operating the plants. For completeness, the energy content of ethylene and acetylene must also be taken into account. The energy requirement for chlor-alkali electrolysis is shared between chlorine and caustic soda. According to Kindler and Nickles, if all these items are added together, one arrives at an energy requirement for the manufacture of 1 t PVC which amounts to 53 GJ. This should be compared to the energy equivalent of polyethylene (70 GJ/t), polypropylene (73 GJ/t) and polystyrene (80 GJ/t).

Processing

Polyvinyl chloride is never processed in its virgin state. In every case a stabilizer has to be added to PVC to protect the plastic from decomposition by heat and light. The most important stabilizers are based on lead compounds (1985: 8,448 t lead). These are predominantly used in the manufacture of cable compounds and construction products. The use of cadmium compounds (1985: 280 t cadmium) is restricted to window profiles, other profiles for outdoor use and roofing sheets. Probably cadmium stabilizers will be replaced within the next few years by stabilizers based on zinc and calcium. The low toxicity of selected tin stabilizers (1985: 478 t tin) and calcium/zinc stabilizers allows them to be used in the manufacture of food packaging, children's toys and medical devices.

PVC is processed at low temperatures around 180 °C, placing low demands on the temperature stability of the pigments. For this reason cadmium pigments are now no longer used for colouring PVC. It is important that the colourants do not bleed. Apart from organic pigments, nickel and chrome titanium pigments, chrome yellow, molybdenum red, iron oxides and ultramarine blue are used. The most important white pigment is titanium dioxide.

Nowadays, one third of the PVC is processed with plasticizers to so-called flexible PVC. This includes cable compounds, plasticized films, coatings, floor coverings and other articles. In 1987 some 160 kt phthalic acid ester based plasticizers were used for PVC processing in the Federal Republic. In addition, there are also other plasticizers based on aliphatic carboxylic acid esters and other compounds.

Impact modifiers are frequently added to rigid PVC. These are other plastics, e.g. ABS, which impart a certain toughness to rigid PVC. Lubricants aid processing. Blowing agents serve to expand PVC foams. The most important blowing agent is azodicarbonamide. Fillers such as chalk are used to adjust mechanical properties. Flame retardants play a secondary

Occurrence of PVC waste and its contribution to the harmful substances content of municipal waste

In 1987 manufacturing industry generated approximately 150 kt PVC waste (calculated as polymer). Of this 60 kt were reused, a further 70 kt were disposed of in municipal waste disposal plants. The remaining 20 kt waste were deposited in company owned landfills or removed to special waste plants. In 1987 domestic waste contained 85 kt PVC, most of it packaging material. Automobile scraps contained about 40 kt PVC. As long-lived building products are the most important use of PVC, reconstruction and demolition of buildings will be the most important source for PVC waste in the future.

In the Federal Republic there is an annual accumulation of around 29 million t municipal waste consisting of household waste, commercial waste similar to domestic waste, bulky waste, road sweepings and market waste. Thirty per cent of municipal waste is incinerated, most of the remainder is landfilled. In 1987, the municipal waste contained approximately 170 kt PVC corresponding to 0.57 % by weight. The PVC content of municipal waste originated from household waste (85 kt), from commercial waste (70 kt) and from other sources (15 kt). Half the chlorine content of municipal waste can be attributed to its PVC content. In 1985, PVC contributed 11 t to the cadmium load of domestic waste. Commercial waste and other sources of PVC waste contained slightly over 20 t cadmium. Therefore 11 % of the cadmium load of municipal waste (a total of 300 t cadmium) could be attributed to PVC. In the same year the PVC fraction of domestic waste contained 46.6 t lead. Slightly less than 800 t lead originated from PVC in commercial waste. In 1985 therefore, PVC contributed about 6 % to the lead content of municipal waste (a total weight of 13,000 t lead). Zinc and tin compounds are used for stabilizing PVC-products for packaging. The nickel and copper content of the PVC fraction is attributable to the use of pigments (nickel titanium or phthalocyanin pigments). Nevertheless the contribution made by PVC to the pollution of municipal waste with zinc, tin, copper and nickel is slight.

Recycling

The same recycling methods used for other thermo-plastics, apply to the reclamation of PVC waste. Recyclates, made by size reduction of clean single material scrap, can wholly or partially replace virgin PVC in many uses. The high density of PVC also makes it possible to reclaim PVC as a single material from mixtures of plastics. Mixed plastics waste containing PVC can be processed to thick walled products. If the plastics are highly degraded or heavily soiled, they can no longer be processed as thermo-plastics. Heating in the absence of air (pyrolysis) or reaction with hydrogen (hydrocracking) destroys the macromolecules yielding products similar to crude oil which can be used as chemical raw materials or as energy source.

Clean, single material PVC waste is comminuted to produce a recyclate which can replace virgin PVC in many uses. At most this requires 5 % of the energy necessary for the manufacture of PVC. This increases to slightly over 10 % of the energy needed for PVC production, if, in addition to size reduction, recycling also requires the scrap to be washed, separated from other plastics in a hydrocyclone, dried and granulated. Processing of PVC containing mixed plastics waste safes the effort of separation and cleaning, but the range of recycling products has to be arranged more attractively. PVC can be processed only a limited number of times before use and processing irreversibly damage the polymer. Hydrocracking converts PVC into hydrocarbons, but the high chlorine content renders this option barely attractive economically. While the chlorine content creates no technical problems in hydrocracking, the splitting off of hydrogen chloride may create technical problems in the pyrolysis of plastics scrap.

It is only the energy content of PVC which is utilized in waste incineration plants. One tonne PVC gives rise to 900 kWh power. In waste incineration PVC causes emissions. By means of PVC, 32 kt chlorine were introduced into waste incineration plants in 1987. Modern waste incineration plants emit about 3 % of the introduced quantity of chloride. Most of the chloride is retained in the flue gas scrubber and has to be neutralized. The resulting salt is either deposited as special waste, or reclaimed as saleable sodium chloride. Incineration of PVC waste is responsible for 0.2 t cadmium emissions, 1.6 t lead emissions and very slight emissions of chromium, nickel and copper. It had been suspected for some time that PVC contributed to the formation of polychlorinated dibenzodioxins (PCDD) and dibenzofurans (PCDF). However, numerous trials showed that the quantity of dioxins and furans which are formed in a waste incineration plant are independent of the PVC content of the waste. On incinerating PVC-free residues, the same quantity of dioxins and furans is formed as occur when incinerating domestic waste containing PVC.

SUBJECT INDEX

Printed in the United States
By Bookmasters